Advance Praise for

CORDWOOD BUILDING
The State of the Art

Rob Roy has combined some brilliant mortgage-free building strategies with the essentials of cordwood construction to the benefit of us all. *Cordwood Building: State of the Art* is a real-world, hands-on helping of personal and professional experience, gleaned from owner-builders around the globe. You won't find a more thorough or up-to-date reference on cordwood masonry…at least not one that's as well-documented and enjoyable to read.
— Richard Freudenberger, moderator of the Continental Cordwood Conference series, and Publisher of *BackHome*

If you're planning to build your own home on land with plenty of trees or near a forest, you should definitely consider cordwood construction. This unique building technique can be easily mastered by anyone who can lift a chunk of firewood, and if you follow the advice in Rob Roy's definitive new book, the result will be easy to build, energy-efficient, and economical. Cordwood houses look really great, too!
— Cheryl Long, Editor in Chief, *Mother Earth News* magazine

Rob Roy has done an excellent job of compiling up-to-date, personal accounts of how others have built their own cordwood homes — starting from the design all the way through to the completion. *Cordwood Building: The State of the Art* is an invaluable tool for anyone considering the construction of their own cordwood home.
— Alan Stankevitz, cordwood adventurer, and author of www.daycreek.com

Cordwood Building: State of the Art is a thorough, articulate and knowledgeable explanation of this unique and earth-friendly building technique. Rob Roy and friends clearly provide all the necessary tools with which to build a cordwood structure. This is a "must read" for anyone considering cordwood construction.

— Richard Flatau, builder and author of *Cordwood Construction: A Log End View*

Cordwood masonry (or stackwall building as we knew it 25 years ago when it first hit the pages of *Natural Life* magazine) is a sturdy, no-nonsense, unique way to build yourself a house. And Rob Roy's new book *Cordwood Building: The State of the Art* is a sturdy, non-nonsense, unique book. Roy, one of the pioneers of the 70s revival, has brought together an impressive array of experts who combine a nostalgic overview of the early days of this unusual style of building with an up-to-the-minute look at current usage and research.

This is a highly useful and inspiring book, which is long overdue.

— Wendy Priesnitz, Editor, *Natural Life* magazine

A most enjoyable read. An excellent blend of history, current practice and hands-on information for anyone interested in this alternative building technique.

— Dr. Kris J. Dick, P.Eng., Department of Bioystems Engineering, University of Manitoba

Cordwood Building is an impressive book. When I talk to people about stack wall building (cordwood), I often tell them they are only limited by their own imagination. It's impossible to explain how good it feels to design and build your own cordwood home. This book provides lots of ideas from many experienced builders across Canada, the U.S.A., and beyond.

— Cliff Shockey, author of *Stackwall Construction: Double Wall Technique*

> 07/16/03
> Hi Aunt Nancy, we finally got some extra copies from the author/editor, Rob Roy... I've signed my chapter. Thanks for all of the positive P.R. Love, —Jim—

CORDWOOD BUILDING
The State of the Art

ROB ROY

NEW SOCIETY PUBLISHERS

Cataloguing in Publication Data:
A catalog record for this publication is available from the National Library of Canada.

Copyright © 2003 by Rob Roy. All rights reserved.

Cover design by Diane McIntosh. Cover photo Rob Roy.
Interior design by Greg Green and Jeremy Drought.

Printed in Canada by Transcontinental Printing.

New Society Publishers acknowledges the support of the Government of Canada through the Book Publishing Industry Development Program (BPIDP) for our publishing activities.

Paperback ISBN: 0-86571-475-4

Inquiries regarding requests to reprint all or part of *Cordwood Building: The State of the Art* should be addressed to New Society Publishers at the address below. To order directly from the publishers, please add $4.50 shipping to the price of the first copy, and $1.00 for each additional copy (plus GST in Canada). Send check or money order to:

New Society Publishers
P.O. Box 189, Gabriola Island, BC V0R 1X0, Canada
1 (800)567•6772

New Society Publishers' mission is to publish books that contribute in fundamental ways to building an ecologically sustainable and just society, and to do so with the least possible impact on the environment, in a manner that models this vision. We are committed to doing this not just through education, but through action. We are acting on our commitment to the world's remaining ancient forests by phasing out our paper supply from ancient forests worldwide. This book is one step towards ending global deforestation and climate change. It is printed on acid-free paper that is 100% old growth forest-free (100% post-consumer recycled), processed chlorine free, and printed with vegetable based, low VOC inks. For further information, or to browse our full list of books and purchase securely, visit our website at: www.newsociety.com

NEW SOCIETY PUBLISHERS   www.newsociety.com

Contents

- Introduction • Rob Roy ix

Part 1: Where We Have Come From

1 • Cordwood Masonry: A Vernacular Architectural Tradition • William H. Tishler 3
2 • My 25-Year Love Affair with Cordwood Masonry • Rob Roy 13
3 • Cordwood Masonry 101 • Rob Roy 21

Part 2: The State of the Art

4 • Stackwall Construction: The Double Wall Technique • Cliff Shockey 37
5 • The Lomax Corner • Jack Henstridge 43
6 • A Round Cordwood House with 16 Sides • Rob Roy 49
7 • Octagons, Hexagons, and Other Shapes • Rob Roy 57
8 • Bottle Designs in a Cordwood Wall • Valerie Davidson 63
9 • Patterned Cordwood Masonry • Rob Roy 71
10 • Electrical Wiring in Cordwood Masonry Buildings • Paul Mikalauskas & Mike Abel 79
11 • Using Cement Retarder with Cordwood Masonry • Rob Roy 85
12 • When It Shrinks, Stuff It! • Geoff Huggins 93
13 • A Mobile Home Converted to Cordwood • Al Fritsch & Jack Kieffer 99
14 • A Shop Teacher's Approach • James S. Juczak 103
15 • Paper-Enhanced Mortar • Alan Stankevitz 109

Part 3: The World of Cordwood Masonry

16 • Stonewood: A Love Story • Wayne Higgins 117
17 • Woodland Treat • Larry Schuth 121
18 • Cordwood on the Gulf Coast • George Adkisson 127

19 • A Cordwood and Cob Roundhouse in Wales • Tony Wrench 133
20 • More Cordwood and Cob • Rob Roy 139
21 • Cordwood in Chile • Hans Hebel 147
22 • One Old and One New in Sweden • Olle Lind 151
23 • Creating with Stone, Wood, and Light • Tom Huber 157
24 • The Community Round House at Pompanuck • John Carlson & Scott Carrino 167
25 • A New Home on an Old Foundation • Stephen & Christine Ketter-McDiarmid 177

Part 4: Go, Thee, and Do Likewise

26 • The Mortgage-free Cordwood Home • Rob Roy 183
27 • Cordwood and the Building Inspector • Dr. Kris J. Dick & Professor A.M. Landdown 193
28 • Cordwood and the Code • Thomas M. Kwiatkowski 203
29 • Cordwood and the Code: A Case Study • John Carlson & Scott Carrino 207
30 • Cordwood Code Issues: Strength and Insulation • Rob Roy 213

• Afterword: Where We Go from Here • Rob Roy 221
• Bibliography 223
• Glossary of Terms 227
• Index 231
• About the Authors 239

Dedication

Cordwood Building: The State of the Art is dedicated to all of the cordwood authors, builders, and innovators who have shared their trials and discoveries so eloquently in these pages and good-naturedly put up with my pestering and nagging. Thank you, my friends.

With great appreciation and no small sadness, I also wish to honor the memory of three of our authors who are no longer with us, except in fond memories. Cordwood builders and authors Tom Kwiatkowski, Allen Lansdown, and Paul Mikalauskas had in common a passion for cordwood and a strong willingness to help others overcome the struggles of building. All three did it with uncommon good grace and humor. Thank you, gentlemen. You are missed but will not be forgotten.

Acknowledgments

It takes a lot of people to make a book like this. Besides all the authors, I wish to thank Chris Plant at New Society Publishers for his faith in the project; copy editor Audrey Keating; art editor Greg Green; Cheryl Long at Mother Earth News for permission to use material from an article called "Mother's Cordwood Cutoff Saw," Mother Earth News (May–June, 1982); cordwood builders Barbara Pryor, Steve Coley, and Jim Rhodes for their photography; Rob Pichelman for his fine illustrations; Catherine Wanek for photos and research used in Chapter 20; Ianto Evans and Lars Keller for helpful suggestions in Chapter 20; Joan Mikalauskas and Helen Cook-Kwiatkowski; and, not the least, my wonderful wife of 30 years, for being a full partner in this cordwood adventure. I couldn't have done it without you, dear Jaki.

Introduction

Rob Roy

Cordwood masonry? To some, this sounds like a contradiction in terms, an oxymoron. Standard wisdom (or rural legend) says, You can't put wood into (on, against) concrete (cement, mortar); the wood will rot. This is the view of the unenlightened, whose experience in building has been limited to using wood on side-grain. In point of fact, you can build strong, long-lasting homes of short logs — called "log-ends" — laid up transversely in the wall within a matrix of mortar. Cordwood construction has been used on both sides of the Atlantic for hundreds of years, as evidenced by the vernacular architecture of Scandinavia, Canada, and the upper Midwest of the United States. In Chapter 1, Professor Bill Tishler tells this part of the story.

Although cordwood masonry seems to have been passed along from generation to generation quite steadily in North America, there was a definite resurgence of interest in the 1970s, spurred on by the independent research of: Jack Henstridge of New Brunswick; the Northern Housing Committee based out of the University of Manitoba; and my wife Jaki and me in northern New York. Our personal story is recounted in Chapter 2.

The resurgence of interest in cordwood masonry has been slow and steady, unlike the flash-fire interest in underground housing in the '70s and the sudden rise in popularity of straw bale building in the '90s. Cordwood's growth is broad-based and on solid foundations. Thirty years ago, there was no how-to information on the subject, just a few references to some of the older buildings. Now, the aspiring owner-builder has a number of books, videos, websites, and

Internet chatrooms to go to for information (all listed in the Bibliography). Our Earthwood Building School has taught cordwood classes continuously since 1979, and others have conducted workshops, on and off, over the past 20 years.

On July 9–10, 1994, Earthwood hosted the first ever Continental Cordwood Conference (CoCoCo), attended by virtually all of the shakers and movers in the field. Five years later, on August 21–22, 1999, the Pompanuck Community in Cambridge, New York held the second CoCoCo. Earthwood published small print runs of both collections of papers that came out of these conferences, but only a few hundred copies made their way out into the world.

Twenty-three of the 30 chapters in Cordwood Building: The State of the Art are rewrites of the best and most useful papers presented at the CoCoCo conferences. As editor for both the original CoCoCo papers and the current work, I contacted the original authors and invited them to rework the articles, checking them once again for accuracy from the vantage point of time, and adding new material when appropriate. The co-operation of these cordwood builders and authors has been nothing short of spectacular. It has been a pleasure to research and re-edit the papers into coherent chapters for this book, a process which often leads to new information. I have written four new chapters and new editorial material, where necessary, to help fill in holes and bind the whole package together. And there are three new case studies appearing in print for the first time in these pages, as well as a new Afterword, Bibliography, and Glossary of Terms.

Part One tells of cordwood's early days and my own involvement in this fascinating building method. Chapter 3, the longest in the book, gives the basic building techniques, to bring the first-time researcher up to speed.

Part Two deals with cordwood building techniques that have evolved over the past ten years or so and that have received little or no attention in previous mass market books on the subject. Cliff Shockey shares his double wall technique for cold climates; "Cordwood" Jack Henstridge tells of an improvement on the "stackwall corners" method; I describe Bunny and Bear Fraser's wonderful method for building a round house within a 16-sided post and beam frame; Val Davidson shares her bottle-end artistry; and Paul Mikalauskas and Mike Abel give valuable electrical wiring tips for cordwood walls. I discuss the use of cement retarder in mortar, and building regularly patterned walls; Geoff Huggins tells of his own successful method of attending to log-end shrinkage; Al Fritsch and Jack Kieffer retrofit a mobile home with cordwood, transforming it into a beautiful and energy-efficient home; and Jim Juczak and Alan Stankevitz tell of their successful experiments with paper-enhanced mortar, sometimes called "papercrete."

Part Three is an exciting tour of cordwood masonry from around the world, including homes built in very hot and humid climates, as well as in very cold climates. Various woods and mortars were tried, tested, and reported on, including Tony Wrench's use of cob as mortar in Wales, an exciting "new" development which may, in fact, take us back to cordwood origins in the far distant past. I follow Tony's chapter with the latest developments in "cobwood," which will be of particular interest to natural building enthusiasts.

Part Four tells you how to own your own cordwood home — as opposed to the bank owning it for you — and provides useful tips for getting through the permitting process, as well as answers to some of the most common concerns of code enforcement officers.

And why don't the mortared-up log-ends rot in the wall? Read on....

Rob Roy, Author/Editor

PART ONE
WHERE WE HAVE COME FROM

1 • Cordwood Masonry: A Vernacular Architectural Tradition 3
   Professor William H. Tishler, FASLA
2 • My 25-Year Love Affair with Cordwood Masonry 13
   Rob Roy
3 • Cordwood Masonry 101..21
   Rob Roy

Chapter 1

Cordwood Masonry: A Vernacular Architectural Tradition

Professor William H. Tishler, FASLA

More commonly called "stovewood" construction a hundred years ago, cordwood masonry represents a relatively unstudied but significant early wood-building tradition in the Upper Midwest as well as in portions of Canada. In Wisconsin, for example, over 70 such structures have been documented, and countless others await discovery by students of vernacular architecture.

With building costs skyrocketing, this unusual system of log construction is enjoying an enthusiastic renaissance as a simple, inexpensive, do-it-yourself building technique. It has been featured in such back-to-the-land journals as *The Mother Earth News*, *Harrowsmith*, *Countryside* and *BackHome Magazine*. The modern renaissance got a boost when Habitat (the United Nations Conference on Human Settlements held in 1976 in Vancouver) featured a cordwood house as a viable low-cost form of shelter for the millions of inadequately housed people around the world. Within a year or two, the Habitat event was followed, albeit unrelatedly, by the publication of books on the subject by Jack Henstridge, Rob Roy, and the University of Manitoba (see the Bibliography).

This chapter will provide a historical introduction to cordwood masonry, with four primary objectives:

1. To describe cordwood as a pioneer building technique.
2. To suggest a method for the classification of such structures.
3. To discuss the geographic distribution of cordwood structures in North America.
4. To provide several theories regarding the origin of cordwood, a subject which has received almost no research or scholarly examination.

Traditional American log construction evolved from the ancient European timber house built of horizontal logs connected at the corners with various interlocking notching systems. Cordwood construction differs significantly in that walls were made from logs cut into short uniform sections and stacked perpendicular to the length of each wall. White cedar wood was usually selected because of its high insulation and decay-resistant qualities, but oak was utilized in the earliest known United States examples built in southern Wisconsin and northeastern Iowa during the mid-19th century. In many instances, the log units were split lengthwise to halve or quarter them into smaller sections. The pieces were then laid up in a bed of wet lime mortar, which encased each chunk of wood but left the cut ends exposed. The resulting wall closely resembled a pile of neatly stacked firewood — hence the terms "cordwood" or "stovewood" construction.

Later in the life of the building, if it was to be used on a more permanent basis or for human habitation, the exposed log-ends were often covered with a masonry coating similar to plaster or with some form of wood siding such as shingles, board and battens or, most frequently, with clapboards. In one unusual example near the Georgian Bay area of Ontario, a veneer of brick was applied over the building.

1.1: This old abandoned cordwood house in Clay Banks, Door County, Wisconsin had been covered with clapboard and rolled roofing at various times in the past. Credit: Bill Tishler

The nature of cordwood masonry suggests several reasons for its practicality. In terms of wood resources, cordwood did not require extensive lengths of straight, high-quality timber necessary for traditional log or timber-framed buildings. In fact, cordwood structures were built in logged-over areas, from small second-growth timber, from trees not harvested by the logging operation because they were too crooked or defective for sawing into boards, and even from the charred wood remnants of burned-over forests. The relative ease of cordwood construction made it an option for unskilled builders. In isolated areas, one man working alone could erect the walls without help from others.

Certainly cordwood buildings could be less expensive to build than more conventional structures. A cost comparison described by a Canadian in 1949 reported that the municipal garage in North Gower, Ontario, when constructed of cordwood in 1937, cost Can$5,000 to build. By contrast, a similar structure built in nearby Stittsville at approximately the same time and to similar specifications, but using conventional block construction, cost approximately Can$14,000. Even today, in Michigan's Upper Peninsula, elderly residents familiar with this construction method refer to it as "Depression building," recalling the hard economic times of the 1930s when the technique was more widely used.

Other advantages of cordwood included: energy efficiency, particularly when air cells were left in the binding mortar; the reuse of available building materials such as abandoned cedar rail fences and the remains of structures destroyed by fire; and the esthetically pleasing exterior facades resulting from such construction.

The pattern of stacked short sections of log remains the basic characteristic of cordwood building, although a number of interesting innovations can be found that were used to add strength and stability to the walls. These characteristics can readily be observed at exterior corners, where the right angle joining of walls presented builders with rather awkward connection problems. Various structural solutions were devised to solve this problem, making it possible to group cordwood building into three basic types.

1.2: This stackwall cornered cottage is located on the Rideau River, County Road #19, outside of Kemptville, Ontario. The photographer was told that it is between 100 and 120 years old. Credit: Wendy Huckabone

The first method of cordwood construction has the structure framed with heavy timbers, in which the wall panels were filled with a nogging of cordwood, resulting in a type of half-timber construction. The timber structural members, usually about 8 inches (20 centimeters) in thickness, determined the length of the flush stovewood sections. Traditionally, utilitarian outbuildings such as barns, sheds, and chicken coops were fashioned in this manner. The full frame technique is found primarily in east central Wisconsin's Door County, a narrow peninsula jutting into Lake Michigan.

The second form of cordwood construction was the solid cordwood wall, built without an encasing structural framework. Several corner fabrication details have been observed. Squared sections of timber were used in a manner similar to cut stone quoins in a masonry wall. Another rather unusual cornering technique was made by crossing alternate layers of cordwood chunks, a detail that can frequently be seen at the end of rows of firewood stacked for the winter. These "stackwall cornered" examples, as they have come to be known, were seldom plastered on the inside and outside or covered with wood siding, and frequently were left completely exposed to the elements. They invariably appear in Wisconsin, west of the Door County Peninsula where some of the earliest cordwood structures exhibit the technique. This was also the method of construction commonly found in Canada.

The third category of cordwood construction incorporated a balloon-frame system of rough-sawn lumber, in which relatively short sections of cordwood were stacked between the studding in the walls. These were placed perpendicular to the run of the wall and were cut

1.3: Corner detail of a stackwall barn in South Gower Township, Grenville County, Ontario; said to be at least 150 years old. Credit: Wendy Huckabone

to match the size of the 2-by-4 or 2-by-6-inch studs, which were spaced from 16 to 26 inches (40 to 66 centimeters) apart. This type of structure was usually built as a dwelling and sided with clapboards, and the cordwood was used primarily for its insulating, rather than its structural, value. Examples of this type of construction appear to have been built somewhat later than the stovewood structures of the two preceding types, but the category seems to have died out with the advent of relatively low-cost and thermally superior insulations made for the purpose.

The technique of building round and other curved-wall cordwood buildings seems to be a modern innovation (beginning, perhaps, around the 1970s), although such buildings may have been an obvious choice in the distant past if anecdotal references to 1,000-year-old cordwood buildings have any veracity. And they may have, since the use of modular masonry units such as stone or cordwood permits structures of almost any shape.

Geographically, the overwhelming majority of America's pre-World War II cordwood structures are concentrated in Wisconsin, with several also located in adjacent states. Two have been observed in Minnesota, and more were built in the St. Paul area by a company specializing in cordwood construction. At least one building, the Norris Miller house (now listed on the National Register of Historic Places), was erected in the Decorah area of Iowa and re-erected by the Norwegian American Museum of Decorah, Iowa in the early 1980s. Several examples have been found in Michigan's Upper Peninsula, a somewhat isolated region where many other structures undoubtedly await discovery and documentation. A single stovewood building has been noted as far west as Montana. In Canada, a search of secondary sources and limited survey work has identified more than 40 examples, only a small fraction of the many cordwood structures believed to exist there.

The origin of cordwood construction remains a mystery, but several hypotheses might be advanced. Unlike log and various forms of timber building brought to America by colonists and settlers, it has been assumed that cordwood fabrication methods might have originated in North America since, until recently, Americans could find no evidence of its use in Europe.

Indeed, with an abundant supply of timber in much of North America, it is possible that the concept of using cordwood for wall construction could have originated in the minds of ingenious pioneer builders who were familiar with firewood, cut and stacked to provide a source of fuel. Not only does stacked firewood seen from a distance resemble a stone wall in appearance, but in some areas, notably a small region in southern Illinois, a traditional form of the woodpile actually resembles a sloping roofed structure. Thus, it is quite possible that cordwood construction may have been one of those spontaneous developments that occurred in several places at the same time as a response to frontier building conditions.

One of the earliest references to this method of building was published in the *Wisconsin Magazine of History* in 1923. The article described an abandoned farmhouse in Walworth County in southern Wisconsin as having walls made of oak that had been felled from the surrounding woods and "sawed and split into sticks fourteen inches in length ... used as nearly like so many bricks as possible."[1] Allegedly completed in 1849 — later research suggests that 1857 was the actual date — its ingenious pioneer builder was a Yankee born in New London, Connecticut who later lived in Palmyra, New York. Could it be that the technique originated there and was carried northward into Canada? Perhaps the diffusion occurred with Loyalists who left the American colonies for Canada during the Revolution.

Could the technique have originated in Canada? Authoritative research on early cordwood building was undertaken some time ago by Wisconsin's eminent architectural historian Richard W. E. Perrin. In his book, *The Architecture of Wisconsin*, and in an article on the subject, Perrin discusses examples of Wisconsin's cordwood buildings, suggesting that their origin "is definitely not European, but very likely Canadian."[2] He bases his conclusion on the cordwood buildings in Quebec and in the Ottawa area of Ontario that have been documented by the Canadian Inventory of Historic Buildings. The method was apparently used in lumber camps, according to Sibyl Moholy-Nagy in her 1957 book, *Native Genius in Anonymous Architecture*.[3] However, Canadian architectural literature seldom mentions stovewood structures, and the method is variously referred to as "log butt," "cordwood," "wood block," and "stackwall" construction. Curiously, the *Encyclopedie de la maison québècoise* even indicates that the technique is "of American influence."[4]

In his book, *An Age of Barns*, Eric Sloane portrays two examples of stovewood structures and suggests a Germanic influence, stating that "the design is best known as that of the German settlers of Wisconsin."[5] This statement is not documented, however, and studies of Wisconsin's German architecture and its Old World antecedents provide no indication of this relationship.

The possibility of a Scandinavian origin for cordwood masonry is suggested in *The Tomten*, a children's storybook about a traditional Swedish fable.[6] The book deals with the adventures of a small gnomelike creature and is handsomely illustrated with watercolor paintings of rural farmsteads and the Swedish countryside by Stockholm artist Harold Wiberg. One full-page illustration portrays a group of early farm outbuildings, including one of obvious cordwood construction.

Numerous inquiries were made of the artist, as well as of architectural historians, schools of architecture, and museums in Sweden for information on the cordwood building technique in that country. Although most of the individuals contacted were unfamiliar with the method, the Swedish Institute of Building Documentation, as well as the Swedish Museum of Architecture provided useful information that included newspaper clippings referring to two examples of cordwood construction in older Stockholm suburbs. One account referred to a two-story "cubicle in Hagalund ... built in 1887 by a blacksmith." Another clipping referred to a second cordwood house built in the suburb of Hagalund by a man who "felled trees, sawed them into pieces of firewood, and began to pile. When he had finished piling and had secured it with mortar, his woodpile consisted of four one-room flats ... the building is of great interest architecturally."

It is interesting to note that officials from both Swedish agencies suggested in their correspondence that the stovewood technique could have been imported from America by returning immigrants. In correspondence with the author, a curator at the Museum of Architecture further stated that cordwood construction using "quality timber ... is not known in Sweden, where good logs would be used in the usual more advanced way. Poor people, however, who could not afford to buy timber have sometimes used firewood in the way you describe. The walls were then hidden behind plaster .... Furthermore, it is reported that workers at sawmills, who could get wasted wood from the employer free of charge, sometimes built their houses in this technique." His letter also indicates that "no research has been made" on this building method.

Evidence of cordwood construction in Norway as well as in Sweden is set forth in an article by Lars Rambøl in the Norwegian journal *Museumsyntt*, entitled "Stovewood Barns: How Did They Originate?"[7] Mr. Rambøl describes the restoration of a stovewood barn at Langsrud in Eidskog and appeals for more information about the method, which is "little known, even among ethnologists." He describes such buildings as "constructed of wood chunks as they appear before they are split for stove-wood. The chunks were placed in layers in wet clay without the addition of any sort of binder (such as hair from a horse's tail or mane, straw, or horse manure, etc.) The height of each stratum ... are approximately 40 centimeters [16 inches], and between each layer there are placed boards as strengthening members. The walls are 35 centimeters [14 inches] thick."

The Rambøl article mentions scattered examples of stovewood barns in Surnadal in NordMøre, and a few places in southern Trøndelag. In his district, Eidskog, "the building technique is well known among older people and had its widest distribution ... in the period from 1890–1920," and that "experts on the other side of the border in Sweden tell us that the stovewood technique had been used already in the 1850s," where it became widely spread throughout the Varmland area.

The validity of a Scandinavian origin was given further support in a 1979 interview with an 80-year-old mason in northern Wisconsin. The builder of several cordwood structures, he confirmed learning this construction method from his father who acquired the technique in his native Sweden before emigrating to America.

Further evidence of a Swedish connection appears in a northern Wisconsin newspaper's article. Entitled "Builds Unique Home on Farm," the article stated that "A Niagara (Wisconsin) man [was] erecting a home like those built on farms in Sweden," and noted that the house "... is fashioned of blocks 8 inches [20 centimeters] long, any thickness imbedded in mortar and will be finished externally with a coating of stucco." Unfortunately, I have misplaced the article, but recall that it was written in the early 1930s.

There is another interesting piece of cordwood masonry history from the early part of the 20th century, regarding the "Formless Concrete Construction Company." In 1926, a building journal contained an article entitled "A Novel System of Construction." It referred to cordwood masonry being used by a St. Paul, Minnesota construction company. The article noted, "The company suggests that this system of construction is particularly adapted to structures containing many curves and odd shapes. It has been used in silos, gasoline filling stations, barns, and conventional dwelling houses." The piece went on to note that "...the cost of this construction is very low as the work can be done by unskilled labor."

Upon researching this matter further, I discovered that in 1921 a patent had been taken out on the "Formless System of Concrete Construction," by one Louis N. Butler. A check of Minnesota census records indicated that Butler was from Wylie, Minnesota, but he later moved to Grand Forks, North Dakota. In 1981, I interviewed the daughter of one of the Formless Company's officers. She had recollections of Butler and remembers the company building a cordwood dwelling. Unfortunately the company never became very active, she indicated, because of the Great Depression.

In 1997, I lectured about stovewood construction in America at the architecture program of Stockholm Technical University. I'd been invited by Inger Norell, one of their scholars researching this method of building. While there, we visited several old stovewood structures in the Varmland region west of Stockholm. Some were in ruins. One was an outbuilding that had been stabilized at an art museum. Another was being renovated into living quarters. The major difference between the Swedish examples and their New World cousins in Wisconsin was that the former used mud rather than lime mortar as the binding material.

Many questions remain unanswered. Is cordwood a folk building tradition that originated in the United States? Can it be given a logical pattern of diffusion into North America? Why did the technique not receive wider acceptance and use? Perhaps growing interest in the economy, ease of construction, and the energy-conservation value of this unusual construction technique — all good reasons for building cordwood buildings again today — will generate more interest in its origins and early use. Cordwood or stovewood construction might then assume a more deserving role in the rich heritage of America's built environment.

(Editor's Note: William H. "Bill" Tishler, recently retired, was professor of landscape architecture at the University of Wisconsin at Madison from 1964 to 2000. Bill and his students were instrumental in having the John Mecikalski General Store in Jennings, Wisconsin listed first in the Wisconsin State Historic Register and later in the National Register of Historic Places. This excellent cordwood building was built in 1899 and was fully restored between 1985 and 1987, thanks to a grant from the Kohler Foundation. It is now open to the public as a museum.

Much of the information in this chapter comes from the author's research and unpublished field notes from 1964 to the present day, including correspondence or interviews with: John I. Rempel of Toronto; Dr. Matti Kaups of the Department of Geography at the University of Minnesota-Duluth; Grover Brinkman, writer and publisher, of Okawville, Illinois; Bengt O. H. Johansson at the Swedish Museum of Architecture; Inger Norell of the Architectural Program at Stockholm Technical Iniversity; and Gustave Bjork, mason, from Wisconsin.

Prior to the Continental Cordwood Conference of 1994, I asked Bill Tishler if he had any further insight into cordwood's origins. He told me that his sense was that cordwood masonry probably developed independently on both sides of the Atlantic by perceptive builders who could see the advantages. There is probably no large-scale geographical "diffusion" situation, although some builders no doubt got the idea from neighbors who had already experimented with the technique.)

Notes

1. See Paul B. Jenkins, "A Stovewood House," *Wisconsin Magazine of History*, 7 (1923), pp. 189–192.
2. See Richard W.E. Perrin, "Wisconsin's Stovewood Architecture," *Wisconsin Academy Review*, 20, No.2 (1974), pp. 2–9; and *The Architecture of Wisconsin* (State Historical Society of Wisconsin, 1967), pp. 27–32.
3. See Sibyl Moholy-Nagy, *Native Genius in Architecture* (Horizon Press, 1957), p. 194.
4. See Michel Lessard and Huguette Marquis, *Encyclopedia de la maison québècoise: 3 siècles d'habitations* (Les Editions de l'homme, 1972), p. 107.
5. See Eric Sloane, *An Age of Barns* (Ballantine Books, 1967).
6. See Astrid Lindgren, *The Tomten* (Coward, McCann, and Geohagan, 1961).
7. See Lars Rambøl, "Stovewood Barns: How Did They Originate?" *Museumsuytt*, 2 (1976), pp. 107–108.

Resources

"Dream Houses Become Reality at U.N. Conference." New York Times, 6 June 1976, p. 24.

Shorter, D.W. Cedar Block Construction, Building Research Division (Ottawa: National Research Council, 1949), p. 2.

CHAPTER 2

My 25-Year Love Affair with Cordwood Masonry

Rob Roy

As I consider the title of this article, I realize that I can't separate my 25-year love affair with cordwood masonry from my 30-year love affair with my wife Jaki. She has been there on every step of the adventure, including the building of four houses and dozens of outbuildings, the formation of Earthwood Building School, and teaching cordwood masonry with me all over the world. Jaki is the quiet partner who keeps this adventure going from behind the scenes, and one of the best cordwood masons around.

Footloose and Fancy-Free

In 1974 Jaki and I left Scotland on a land search in the United States. Our intent was to find a place where we could pursue a self-reliant lifestyle — tough to do at our little cottage in the Highlands. Jaki was British. I still had US citizenship, although I'd been living in Scotland for several years. During our 100-day search for land, we stopped for 10 days at an intentional community of owner-builders — the structures were domes and log homes mostly — in Arkansas. We were soon offered a job laboring on a log home being built by some of the community members for a client named Joe. This was a great opportunity, as it was part of our plan to build our own home, and a horizontal log cabin was a common building style amongst back-to-the-landers. Not much was known (and even less published) about cob, straw bale, underground, and other alternative building styles.

After a week of laboring on Joe's home, Jaki and I came to the conclusion that building a home of logs large enough to keep the winter cold out would be very difficult, at least in Northern New York, where we had already put a deposit down on 64 acres (25 hectares). Logs of sufficient size would be both expensive and extremely difficult for us to heft.

Serendipity stepped in. Joe, our employer, subscribed to the then-young *Mother Earth News* ("More than a magazine; a way of life"). He also had the current (April, 1974) *National Geographic* magazine, with a photograph of a cordwood home built by a woman and her son in Washington.

Later in 1974, we returned to Scotland to sell our cottage, the proceeds from which would grubstake our homesteading adventure. During this time, we also found a reference to cordwood masonry in Eric Sloane's book, *An Age of Barns* (Ballantine Books, 1967). Sloane called the building technique "stovewood" construction, and we soon learned that many Canadians called it "stackwall" building. Whatever we called it, Jaki and I knew right away that cordwood masonry was a building system we could do ourselves, alone. Plus, it made sense from a fuel efficiency standpoint because walls could be made virtually any thickness, with 24-inch-thick (60-centimeter-thick) walls not uncommon in Canada.

We moved to our property near West Chazy in northern New York in late April 1975. Thanks to correspondence resulting from a "Positions and Situation" listing in *Mother Earth News*, we were joined by several other would-be homesteading families, and the Murtagh Hill community was born. I am happy to report that it is still going strong, and the second generation has begun to build.

Log End Cottage

We bought our land from an old North Country builder, Tom, who seemed to take a personal interest in us and wanted to help us launch our adventure. Tom helped us frame out our temporary shelter, a 12-by-16-foot (3.6-by-4.8-meter) shed, and also took us up to Canada's Ottawa Valley, just two hours' drive away, to search for cordwood homes. So, on a Sunday afternoon in the spring of 1975, we found several old (probably 19th century) cordwood homes still occupied and, on the way home, had the good fortune of coming across a farmer building a cordwood barn.

We stopped and spoke with the gentleman for a half hour or so, hanging on his every word of advice, as we had been unable to find any sort of how-to instruction in any of the few references we'd found so far. "Use plenty of extra lime," he advised. "Extra lime," we noted dutifully. Just to see that barn going up told us that here was a good building method, one, as we had learned earlier in the day, that would last a long time, and one that we could easily manage ourselves.

We wanted a home that would be visually pleasing and have a certain old world atmosphere to it. You know — exposed heavy beams, that sort of thing. We decided on a Swiss chalet design, with plenty of overhang to protect the cordwood masonry. We scoured the area for recycled heavy barn timbers for a post and beam frame, as we had no great faith in cordwood masonry as load-bearing in those early days, and we were thinking of putting a heavy earth roof on the home. Now, we know that cordwood *can* bear very heavy loads.

We built the frame first, got the roof on, and enjoyed the compelling advantage of doing the masonry under cover, protecting the mortar and insulation cavity from sun and rain. We still believe strongly in the advantages of building within a post and beam frame, although Earthwood, the house we have lived in for over 20 years, has load-bearing cordwood walls with a very heavy earth roof.

2.1: Log End Cottage, the author's first cordwood home, built in 1975–76 in West Chazy, New York.

Log End Cottage, as it came to be known, was a charming romantic bungalow, esthetically and romantically suited to its still freshly married builders. We made plenty of mistakes, and we even had one or two successes, including the discovery, on the very last panel of the house, that sawdust could be used to retard the set of the mortar, thus preventing mortar shrinkage cracks. We refined our mix over the next few years and arrived at the mortar recipe that we now teach at our classes (see Chapter 3).

Our mistakes were several and are worth relating. My father was fond of saying that a smart man learns from his mistakes, and a wise man learns from the mistakes of others. Here are our major goofs.

1. *Wall thickness.* The walls were two narrow for our North Country climate of 8,600 degree days (see Chapter 3). To match our barn timber framework, we decided on a wall thickness of 9 inches (23 centimeters). While the cottage had charming comforts of its own, energy efficiency was not one of them. In the three years that we lived there, we consumed an average of seven full cords of hardwood each winter. Too much.

2. *Basements.* The basement was not a success. We put half of our $6,000 expenditure into a full basement, which space was probably used less than five percent of the time. Since Log End Cottage, I have not spoken well of basements, although I *am* in favor of high-quality, earth-sheltered space — a totally different concept — as we achieved at our second home, Log End Cave.

3. *House shape.* Log End Cottage was twice as long as it was wide, yielding a poor relationship between perimeter walls and enclosed space. Such a shape, typical in the ubiquitous American ranch house so popular in the '60s, has a poor ratio of useful

space per unit labor and materials. Birds, bees, and beavers know instinctively and without a course in geometry that a round structure makes the most efficient use of labor and materials. But most human builders (at least in the Western world) prefer square corners — probably out of habit, but also because of the availability of lumber, sheet materials, and rectangular masonry units. If you absolutely must have a rectilinear house, may I suggest that the closer to a square you make your design, the cheaper the per-square-foot (per-square-meter) cost, and the sooner the home will be completed. In our 25 years of experience, Jaki and I have developed efficiencies for the handling of materials, which we share with our students and through our videos, but the bottom line is still that cordwood masonry is labor intensive. There is no getting around that, unless you choose to compromise quality, which we do not advise.

4. *Poor orientation.* The Greeks knew thousands of years ago that the orientation of any home, sensibly or poorly conceived, can make a 35 percent difference in energy efficiency. A town near us had a zoning ordinance that required homes be built parallel to the road, without regard to the direction that the road ran! Such an ordinance, no matter what the other energy qualities of the house might be, could easily subject homeowners to 35 percent higher energy costs. Log End Cottage ran north-south; an east-west orientation would have greatly increased solar gain in the winter.

Log End Cave

2.2: Log End Cave, built by Rob and Jaki Roy in 1977.

We fixed our original errors in our second home, Log End Cave, but made some fresh ones. My father, full of great quotes, said, "You have to build two houses to get one right." Well my father was obviously brighter than me, for it has taken us three tries to get one really right. Still, Log End Cave, built just 150 feet (45.7 meters) from the Cottage, was a huge improvement in what I call "livability." Although most of the home was earth sheltered, the part above grade (mostly on the south side) was of cordwood masonry, 10 inches (25 centimeters) thick this time — an 11 percent improvement. Large south-facing, double-paned insulated windows made for good solar gain in the winter, and the earth-sheltered aspect kept the house a nice, steady, comfortable temperature. And despite being a so-called underground house, the Cave was much brighter than the Cottage. Although Log End Cave had

20 percent more useful space than the Cottage, the fuel requirement dropped to three cords — just 43 percent of the amount burned at the earlier home.

There was no basement. Rather, we went the extra mile to incorporate good drainage, waterproofing, and insulation to the exterior of the 12-inch-thick (30-centimeter-thick), surface-bonded concrete block walls below grade. The interior part of the exterior walls were a pleasing textured white, which looked good and promoted plenty of light. Three roof skylights also helped a great deal in making the Cave a much brighter home. It must be stated somewhere — so why not here? — that cordwood masonry, like stonework, is a light-absorbing surface. It is a light sucker. Therefore, we advise that interior walls be white or at least very light in color, to reflect light back onto the cordwood walls.

2.3: Log End Sauna, built by the Roys in 1979, is still used as a sauna today.

And there were new errors. There was only one entrance into the home, on the south side. This is not only a code violation, it is also unsafe and stupid. A chipmunk has more sense. If a fox comes to the front entrance, the rodent can skip out the back. Maybe the code enforcement officer comes to the front door ....

Another error was a lack of opening windows. Every room should have opening windows. We did have large, screened vents in each room, but these would not have provided escape, if necessary. We put in plenty of opening windows at Earthwood.

Finally, although we did insulate Log End Cave with rigid foam insulation, correctly on the exterior of the mass (the 12-inch concrete block walls), we should also have insulated around the footings and under the floor with an inch of extruded polystyrene. In the late spring and early summer, condensation would occur where the internal wall meets the floor. This was eliminated at Earthwood by insulating around the entire fabric of the building, thus keeping interior surface temperatures above dew point. These errors were not cordwood errors, but many people combine cordwood masonry with earth-sheltering, so I feel that they are worth mentioning here.

We lived in Log End Cave, a bright and comfortable home of 910 usable square feet (84.5 square meters), for about three years. The basic home cost $6,750. We knew that we were ready for our dream home, and in 1981 we began construction of Earthwood House, a half-mile up the road. We had already begun teaching cordwood masonry classes, but we did not call ourselves Earthwood Building School until 1980. Students helped us to build Log End Sauna in 1979, still in use as a sauna today.

Earthwood

Taking a leaf from the notebook of underground guru and architect Malcolm ("Mac") Wells, we chose to build Earthwood in an abandoned gravel pit. Mac says that we should not be building houses on the best land that nature has to offer. Better that we take marginal land that has, for example, been clearcut or gravel-pitted or otherwise diminished and help restore it. Now, 20 years after building Earthwood, we are proud that we have returned almost 2 acres (0.8 hectares) of dead lifeless moonscape to green, living, oxygenating production. We have done something positive for the planet. Another advantage of this strategy is that you can purchase such marginal land at just a fraction of the cost of pristine land.

Earthwood is a round, two-story, earth-sheltered cordwood masonry house with an outside diameter of 38 feet, 8 inches (11.8 meters). Sixty percent of the walls are above grade and consist of 16-inch-long (40-centimeter-long) white cedar cordwood, windows — lots that open! — and doors. The rest of the cylinder is below grade and is constructed of surface-bonded corner blocks (8 inches by 8 inches by 16 inches; 20 centimeters by 20 centimeters by 40 centimeters), laid transversely in the wall like cordwood, giving a 16-inch-thick (40-centimeter-thick) wall. The footings and downstairs floor are concrete, all insulated between the fabric of the building and the earth with 1 – 3 inches (2.5 – 7.6 centimeters) of extruded polystyrene. (Dow Blueboard™ has the best compression strength for use under footings.)

The roof is covered with 7 inches (18 centimeters) of earth. Below the earth is a 2-inch (5-centimeter) crushed stone drainage layer over 4 inches (10 centimeters) of Dow Blueboard™. The insulation rests right on the W.R.Grace Bituthene™ waterproofing membrane, which is directly installed over the tongue and groove plank roof.

Earthwood was a big project. It took us seven months to get the building closed in to the point where we could heat it, and six more months to complete the interior work. We were in no hurry to move into a construction site, as we had the comfortable Cave to live in close by. (See the Color Section)

We didn't make any lasting mistakes at Earthwood, although we did learn something extremely important about cordwood masonry. Our previous homes had been built of eastern white cedar, a great choice of wood because it is so stable when dry. I wanted greater mass below grade, and tried to go with bone-dry split hardwoods there. This was a mistake. The wood took on moisture from rain collecting on the slab, swelled, and caused the wall to tilt out 3 inches (7.6 centimeters) at 6 feet (1.8 meters) of height — an unacceptable situation with another story and a heavy earth roof still to come.

We tore the wall down, tried rebuilding with expansion joints every 8 feet (2.4 meters) — too little, too late — and tore the wall down again. For five weeks we built walls and tore them down again until we had a real, clear perception that you don't want to build cordwood walls out of very dry hardwoods. We fell back to the old tried-and-true method of surface-bonded concrete blocks below grade, and this is what we now recommend to anyone wanting to build an Earthwood-type house.

Earthwood has been a great success, and cost us less than $10 per square foot ($32 per square meter). It feels great living in a round house that provides an almost womblike comfort. The 2,000-square-foot (186-square-meter) house is cool in summer and warm in winter, and is easy to keep in the low- to mid-70s Fahrenheit (low- to mid-20s Celsius) all year round on just 3.25 full cords of firewood per year. This exceptional performance is the result of five different design features: 1. We used 16-inch (40-centimeter) insulated cordwood masonry walls above grade; 2. The home is earth sheltered — mostly the northern half of the house; 3. We made use of good solar orientation; 4. The round shape minimizes the amount of surface area subject to heat loss; and 5. Most of our firewood is burned in a highly efficient 23-ton (20.8 metric ton) masonry stove, also known as a "Russian Fireplace."

The Earthwood house is now the center of our Earthwood Building School campus, which has several other cordwood buildings, including a garage, two guesthouses, office, library, sauna, and playhouse.

2.4: The Earthwood office is 20 feet (6 metres) in diameter.

2.5: During the early '90s, the author and his wife built their "Mushwood" summer cottage at Chateaugay Lake, New York.

Mushwood and Beyond

During the 1990s, on a pay-as-we-could-afford basis, Jaki and I — and Earthwood students — built our Mushwood summer home at Chateaugay Lake, New York. The home consists of a 22-foot (6.7-meter) diameter lower story with two bedrooms and a bathroom, and an open plan 29-foot (8.8-meter) diameter upper level dome, containing the living, dining, and kitchen areas. A curiosity of this design is that the upper level has twice the useful floor area as the downstairs. The cordwood walls are 12-inch-thick (30-centimeter-thick) cedar rounds, with lots of

2.6: A cordwood masonry workshop conducted by Rob and Jaki Roy near Bakersville, North Carolina.

special design features such as mushrooms, shelves, and bottle-end designs. The dome? We love it, but would never build another. It is true that the framework goes up in a day, but then it takes forever to finish the roof and the interior. It is a lovely bright space, however.

I will not dwell on the Mushwood design. We like it but would not encourage inexperienced owner-builders to attempt it. Actually, inexperienced owner-builders should be wary of starting with a house the size of Earthwood, too. While there is nothing intrinsically difficult about the Earthwood construction or techniques, it was simply a huge project. It put a strain on our relationship — and we had the advantage of a few years of building experience! Happily, the strains we felt never took us to a danger point in our marriage, although I have known several cases where the traumas of long owner-building projects have certainly been a contributing factor in relationship breakdowns.

Jaki and I continue to build and teach cordwood all over the world. The projects belong to others, not us, but we still get emotionally tied up in each one, wherever and whatever it may be. You see, cordwood masonry is fun, almost intoxicating. And you know the best part about it? The people. Cordwood people, who have already demonstrated the ability to "think outside of the box," are about the nicest folks you can meet. They are creative, industrious, intelligent and, most of all, have a sense of humor.

We are fortunate to make a living out of our love affair with cordwood and most fortunate of all to enjoy our work with such fine people.

(Author's Note: For further information on building a Log End Cave type earth-sheltered home, see The Complete Book of Underground Houses *(Sterling, 1994). For details on building the cordwood saunas at Log End or at Earthwood, see* The Sauna *(Chelsea Green, 1996). The step-by-step construction of Earthwood is detailed in* Complete Book of Cordwood Masonry Housebuilding *(Sterling, 1992). For a much more thorough account of our own personal journey, see* Mortgage Free! Radical Strategies for Home Ownership *(Chelsea Green, 1998). All are available from Earthwood Building School at: www.cordwoodmasonry.com.)*

CHAPTER 3

Cordwood 101

Rob Roy

Cordwood Building: The State of the Art is full of new case studies from around the world and new technical innovations since the publication of *Complete Book of Cordwood Masonry Housebuilding* (Sterling, 1992). But this new material must be built upon a solid foundation of basic cordwood masonry theory and techniques. The intent and purpose of this chapter, then, is to bring the cordwood masonry novice up to speed.

The Wood

To build a cordwood home, you first need ... cordwood! See, we really are going to start with the very basics. Over the years, I have found that seven questions come up time and again about the cordwood itself. We'll look at them one at a time.

1. *What kind of wood is best?* The best choices for cordwood building are the more stable species — that is, the kinds of woods that shrink and expand least. The problem that occurs most often is shrinkage. However, log-end shrinkage, while irritating, inconvenient, and esthetically disturbing, is not a critical problem. There are things that can be done about it (see Chapter 12). Wood expansion, however, while much rarer, *can* be a critical problem, as we found out when we tried to build the back wall at Earthwood with very dry hardwood log-ends.

    When wood wants to expand, there is nothing we can do to resist it. Granite quarrymen in the 19th century would drill several ¾-inch holes behind the face of granite they wished to split. They would insert dry, ¾-inch hardwood dowels (such as oak), water them, and after a while, the swelling oak dowels would break off an 18-inch (46-centimeter) face of granite! With a curved cordwood wall, this wood expansion will cause the wall to go out of plumb. At Earthwood, despite careful

3.1: Steve Coley removes bark with a drawknife. The tops of the posts are notched, and the rear post has a stop carved into it to provide resistance, thus holding the log steady. Credit: Barbara Coley.

building, the expanding hardwood log-ends sent the wall 3 inches (7.6 centimeters) out of plumb at 6 feet (2 meters) of height! With post and beam frameworks, the expanding wood can push corner posts out (no matter how they are fastened down) and/or cause uplifting of plate beams at the top of the cordwood wall. Stackwall corners, made up of alternating corner pieces called quoins (or Lomax units, as described in Chapter 5), will be pushed out in both directions by expanding cordwood. Although rare, wood expansion is a critical problem that must be avoided.

Expansion and shrinkage are related. In general, woods most prone to shrinkage are also the ones most prone to expansion. The more stable woods are what I call the light and airy woods, such as white cedar, larch (tamarack), white pine, spruce, cottonwood, lodgepole pine, quaking aspen, and the like. These woods can be used fully dry without real serious expansion problems. And if they *are* dry (a year or more at log-end length), they will shrink very little. Red pine, and Virginia and red cedar have been fairly successful. Hemlock is prone to great shrinkage.

Hardwoods, such as oak, maple, birch, beech, and elm, as well as some dense southern pines have potential expansion problems, particularly if they are dried too long before building. I don't know all of the woods that are out there in different parts of the world, but in general, you are looking for lightweight airy woods, not dense, heavy, fine-grained woods, which tend to both shrink and expand a lot. Look for local woods with low shrinkage characteristics. Also, airy woods have a better insulation value than the dense hardwoods.

Rot resistance is not as big a factor in wood species choice as one might expect. Wood rot is caused by fungi, which need nutrients, air, and constant moisture to propagate. With a cordwood wall, the little varmints have only the first two requirements, not the third. Because log-ends are constantly breathing along end-grain, moisture is never trapped. (See the sidebar on the five things to do to prevent wood rot.)

2. *How long should the wood be dried?* It depends on the kind of wood. With the more favorable light and airy woods, there is generally no problem drying the wood a year or more. A year's drying at log-end length will go a long way toward preventing shrinkage with these woods, and expansion should not be a problem. If you must use the denser species of wood because that is all you have on your land or in your area, just split and dry the wood for six weeks or so. Yes, there will be shrinkage, but this can be taken care of a year or two down the road (for methods, see Chapter 12). It really isn't worth taking a chance with expansion.

> ### Five Rules to Prevent Wood Rot in a Cordwood Wall
>
> Log-ends, because of their breathability, are not prone to deterioration in a cordwood wall in the first place. And if these five rules are followed, the chance of wood rot will diminish to almost nothing.
>
> 1. Keep the cordwood masonry elevated at least 4 inches (10 centimeters) off the ground on a good concrete block, concrete, or stone foundation. In wet climates, up this to 1 foot (30 centimeters).
> 2. Use a good roof overhang all around the building. I like a 16-inch (40-centimeter) overhang, but 24 inches (60 centimeters) or more is even better.
> 3. Don't allow two adjacent log-ends to touch each other or a surrounding post and beam frame. Moisture can get trapped there and promote the growth of fungi.
> 4. Build only with log-ends that are sound in the first place. Reject wood that shows any sign of existing rot or deterioration.
> 5. Debark the wood. Insects love to get between bark and the outer layers of the wood.

Incidentally, wood dries ten times faster on end-grain than through its outer layers. Therefore, the real drying takes place after longer logs are cut into their final log-end length. If you don't see a split (called a "check") on the outside of a 10-foot-long (3-meter-long) log that has been "lying around for three years," the chances are it is still going to do a lot of drying and checking after it is cut into short pieces. Split wood also dries faster than unsplit wood. Dry the wood in single ranks, kept off the ground on wooden stringers or pallets. Cover only the top of the rank, not the sides. Covering the sides will create a greenhouse effect, trap moisture, and make the rot-producing fungi very happy indeed.

3. *Should I bark — or "debark" (means the same thing) — the wood?* Yes, definitely. The space between the bark and the epidermal layers of the wood is a great place to trap moisture and provide habitat for unwanted little houseguests. It is easiest to bark the wood in the spring when the sap is rising, hardest in late autumn. When it is easy, almost any sharp or flat tool will serve as a peeling spud: an ax, pointed trowel, scraper, or even a garden hoe that has had its business end straightened flat. When barking is difficult, the tool of choice is a drawknife. Cordwood builders Barbara Pryor and Steve Coley of Amherst, Virginia found it very worthwhile to shape tree

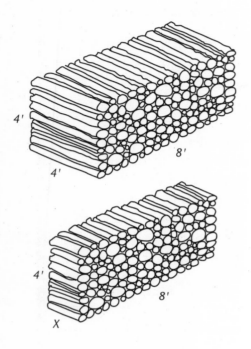

3.2: Above is a full cord, below is a face cord, where "X" is whatever length of wood the seller is supplying: 12 inches (30 centimeters), 16 inches (60 centimeters), etc.

stumps into bracing supports of convenient (non-backbreaking) height. Jaki and I found drawknifing — normally a killer of a job — to be a pleasure with the long logs supported in this way.

Years ago, in the spring, neighbors cut a large quantity of cedar for a log cabin. When the wood was first cut, the bark came off easily, but our neighbors were into a production mode, so they decided to wait until all their logs were cut before commencing serious debarking. When, after a couple of weeks, they returned to the early logs, they found that the nice greasy layer of sap that normally makes the bark so easy to peel had turned to glue, and they had to remove the bark with a drawknife —a much more tedious process. Cut your trees when the sap is rising — experiment to find out when this is, starting in very early spring — and peel the bark off the trunks and limbs before the wood stops vibrating from hitting the ground.

In a worst-case scenario, you can adopt a method discovered by Steve and Tara Myers of Tonganoxie, Kansas (see Mission Impossi-peel: How We Barked Our Logs and Kept Our Sanity).

4. *How much wood should I cut?* This is a good question and it gets very little coverage in the literature. The best measure to work in is — no surprise — the cord. Now, a real cord, a *full* cord, is actually a stack of wood 4 feet wide by 4 feet long by 8 feet long (1.2 meters by 1.2 meters by 2.4 meters), or 128 cubic feet (3.6 cubic meters). But full cords and cubic feet confuse the issue. The calculations are easier and more accurate if we work in something called "face cords." Face cords are also 4 feet high and 8 feet long. But the depth or thickness of the stack is whatever uniform length the wood is cut: 12 inches (30 centimeters), 16 inches (40 centimeters), 24 inches (60 centimeters), or whatever. So the area of the side of a face cord is always 32 square feet (3 square meters), and this is the magic number we can use in our calculations.

From your plans, figure the square footage (square meterage) of wall area which is actually cordwood masonry. Subtract doors, windows, and heavy timber framing from the gross wall area to arrive at this figure. A house with a perimeter of 125 feet (38 meters) and a height of 8 feet (2.4 meters) has 1,000 square feet (91 square meters) of wall, gross. Say the windows, doors, and post and beam frame make up 20 percent of the wall. (You can figure this accurately from your plans.) Subtracting 20 percent — 200 square feet (18.2 square meters) in this case — leaves 800 square feet (74 square meters) of actual cordwood masonry. Now divide by the magic number 32 (3),

Mission Impossi-peel:
How We Barked Our Logs and Kept Our Sanity
Steve and Tara Myers

Like all soon-to-be cordwood builders, we were anxious to get our cedar cut, stacked, and peeled before the summer arrived. So we meticulously felled seven to eight cords of wood and cut it to length with our tractor's buzz saw. As we put the logs through the saw, we attempted to begin the process of debarking, but the bark held fast to the wood like a magnet. Research into the subject told us that the bark might be easier to peel after drying it for a while. Well, we waited ... and waited ... and waited. We'd go out to the woodpile every couple of weeks to check our PQ: peelability quotient. It wasn't improving and we were starting to get a bit testy.

After several sleepless nights, we decided to try a last ditch experiment. We'd noticed that in the process of moving some large uncut logs off the ground to prevent them from rotting, the bark (where it was wet) virtually fell off the log. So we immersed several short logs in a bucket of water and checked daily. After three days, the bark was loose enough to peel with a knife. Yahoo! We then proceeded to fill several stock tanks with water and log-ends cut to length. This method worked well for us because we peeled on weekends. The tanks held just enough cordwood to keep us busy peeling and stacking for two days. The wood remained wet on the outside for a short period of time but returned to its previous state quite quickly, so drying time was not adversely affected.

We hope this information will help others in their efforts to peel their cordwood. It may not be the quickest method, but it is effective for wood that would otherwise be impossible to peel!

(Editor's Addendum: Thanks for the tip, Steve and Tara. Here's another for really tough situations. A chainsaw attachment called Log Wizard is manufactured by Goldec International Equipment of Red Deer, Alberta. This device adapts to both .375-inch and .325-inch chain and allows your saw to be used for debarking, post sharpening, or as a notcher/planer. Call Goldec at: (403) 343•6607 to find out where to purchase the Log Wizard near you. It can also be purchased from Bailey's, a logging supply company. Call (800) 322•4539 for a catalog.)

which yields, in the example, 25 face cords. Finally, you can safely discount 20 percent from of this number, because the cords swell by at least that much when they are restacked with mortar. So if you had 20 face cords laid by, at whatever length of log-end matches the thickness of your wall, you will have plenty of wood, enough to reject misshapen pieces that you don't like or that are troublesome to use.

Degree Days

The annual total of degree days for a heating season is a useful yardstick for comparing fuel requirements for different climates. It works like this: The degree days (DD) for any day in the heating season (generally September to June inclusive, in the north) is the difference between that day's average temperature and 65 degrees Fahrenheit, the baseline comfort level. For example, if the average temperature on a December day is 20 degrees Fahrenheit, the DD for that day is 45 DD. (65 − 20 = 45). Adding up all of the individual DD totals for the heating season gives the annual DD total. In Canada, 18 degrees Celsius is the baseline, so a conversion must be made before drawing comparisons.

Where we live, an hour's drive south of Montreal — and 1200 feet (366 meters) in altitude — the average season has about 8600 DD. Washington, DC, by comparison, has about 4300 DD per average season. Any particular house — energy efficient or inefficient — will require about twice as much fuel to heat near Montreal as near Washington, DC. The numbers work quite well that way. A particular winter could vary quite a bit, perhaps as much as 10 percent, from the long-term averages upon which the published figures are based. If we have a mild winter of, say, 7740 DD (10 percent less than normal), then we find that we use about 10 percent less fuel during that season, about nine face cords of 16-inch (40-centimeter) firewood instead of the usual ten. These comparisons work for any measure of fuel.

Thirty-year average heating degree day figures for the United States are available at: <www.nws.mbay.net/hdd.html>. For cooling degree days — of interest for those in the southern United States — go to: <www.nws.mbay.net/cdd.html>.

---

5. *How thick should the walls be in a cordwood home?* There are lots of unspoken variables in this question: climate (degree days), type of wood, shape of house, etc. We are pleased with the performance of our 16-inch (40-centimeter) white cedar walls at Earthwood, in a heating climate of 9000 degree days (see Degree Days). Our 16-inch white cedar cordwood walls have an insulation value of about R-19 or a little better. In Canada and Alaska, 24-inch-thick (60-centimeter-thick) walls are quite common and make sense. In the South, where the energy cost of cooling can equal or exceed the heating cost, 12-inch (30-centimeter) walls are adequate, but the thermal mass of thicker walls might also help to make the home even easier to keep cool. George Adkisson tells me that the 12-inch-thick cordwood masonry walls of his home on the Gulf Coast of Texas reduce his air-conditioning costs to about half that of similarly sized conventional homes in the area (see Chapter 18).

6. *How should I cut the wood?* Most people use a chainsaw to cut long logs into log-ends, and that is how I do it as well. I simply mark the long piece with a lumber crayon every 8 inches (20 centimeters) or 16 inches (40 centimeters) or whatever size I want. Then I destroy the crayon mark with my chainsaw cut. The saw cut will take about a ¼-inch (6-millimeter) "kerf" out of the wood, so a 16-inch (precisely 40.6-centimeter) log-end will actually be about 15¾ inches (40 centimeters) long, and I figure that a ½-inch (13-millimeters) either side of this is acceptable. If you want really precise lengths, and a safe and easy method of cutting, make yourself a cordwood cutoff table for your chainsaw (see Chapter 9).

   Another good way to cut cordwood is with a large circular saw, typically 30 inches (76 centimeters) or so in diameter. These saws are commonly connected to a tractor's "power take-off" (PTO), by way of a belt. The long length of wood is set on a movable table. The table, with the log on it, is tilted toward the saw, which cuts the ends off quickly with a nice straight cut.

   Cutting log-ends by any means must be considered a dangerous activity. Use proper ear and head protection. Logger's safety chaps are a good idea for leg protection. Keep all animals and unnecessary people, especially children, away from the cutting area. Be careful. Your safety is up to you. Before using any kind of cutting equipment with which you are unfamiliar, get training from an expert.

7. *Split wood or round log-ends?* The main reasons for splitting wood are to accelerate the drying process, to eliminate the large "primary check" seen in the rounds, and to reduce the size of shrinkage gaps. Shrinkage is proportional, so the smaller the log-end, the smaller the shrinkage between wood and mortar. But smaller pieces require more handling of materials, and mixing more mortar, too.

   Jaki and I have built beautiful cordwood walls of all split wood, all rounds, and a mixture of the two. In fact, as I completed the previous sentence, I asked Jaki which she preferred. We discussed it for a few minutes and agreed that all styles can look good, and that we really didn't have a preference. The important thing is to maintain a consistency of style, which means making a conscious effort to deplete the various sizes and shapes of log-ends in your stock at the same rate. Which is better? If you have a strong personal preference, go with it.

3.3: *All rounds. All splits. Splits with featured rounds.*

All rounds   All splits   Splits with featured rounds

The Mortar

Lots of different cordwood mortars have been used successfully in different climates, so I don't want to be dogmatic about this. Over the first six years of our cordwood masonry experience, Jaki and I gradually refined a mortar mix that has worked very well for us, and we continue to teach it at our classes. This mix makes use of sawdust as a cement retarder. A mortar that dries slowly will shrink less or not at all, eliminating those irritating little mortar shrinkage cracks between log-ends.

The problem is that suitable sawdust is not always available, and there is always a little bit of doubt about whether or not the sawdust is right for the job. "Suitable" sawdust, in our experience, has consisted of the larger and less dense particles of softwood sawdust that come from a sawmill where logs are made into lumber. White cedar, white and red pine, spruce, and even poplar have worked well. Oak and other dense hardwood sawdusts have not proven to be successful. The hard little cubes of oak do not seem to hold and store the moisture the way that the softer, lighter softwood sawdusts do, and mortar shrinkage is the result. In fact, the hardwood sawdust seems to make the mortar more grainy, crumbly, and harder to point. If you cannot get suitable sawdust or are unsure, Jaki and I strongly recommend the use of one of the commercially available cement retarders (see Chapter 11). But if you do have good sawdust, here are two mixes that have worked well for us. The first makes use of Portland cement, and the second uses masonry cement. The proportions given are equal parts by volume.

Portland Mix
- 9 parts sand
- 3 parts soaked sawdust
- 2 parts Portland cement
- 3 parts hydrated lime

Masonry Cement Mix
- 9 parts sand
- 3 parts soaked sawdust
- 3 parts masonry cement
- 2 parts lime

We have found that these two mixes are very similar in terms of hardness, strength, workability, and smoothness. The main difference is in color. The Portland tends to be lighter and more of a green-gray, whereas the masonry mix is more of a blue-gray. But even these generalities can fall apart when different brands of cement are used. There is a product in our area called "light masonry cement," for example, that yields a very light, almost white mortar, much like Portland.

The sand you use should be washed masonry sand, not the coarse-grained sand used for drainage. You may have to pay more for the finer-grained masonry sand (which has the texture of brown sugar) than for the coarse stuff, but it is worth it. Coarse sand yields crumbly mortar, frustrating to work with. Also, the color of the sand will affect the color of the mortar. Light-colored sand gives light-colored mortar. Dark sand, dark mortar.

The sawdust should be the softer, lighter type, as already discussed. Plus — and this is very important — it should be passed through a half-inch screen and thoroughly soaked at least overnight in a non-leaking vessel, such as an open-topped steel drum or an old bathtub. So, the last thing to do each day is to make sure that enough sawdust is soaking for the next day's work. Portland cement, Type I or Type II, is full-strength cement. You can be sure of its strength characteristics, as long as the powder hasn't gotten damp and begun to set before it is to be used.

Masonry cement can be great stuff, too, but there are a variety of types with slightly different characteristics. I've had good luck with Types M and N masonry cement. Masonry cement has characteristics of Portland mixed with builder's lime. Many masons use a mix of 3 parts sand and 1 part masonry cement as a good all-purpose mix for a variety of masonry applications. (There is further commentary on masonry cement in Chapter 11.)

The lime is builder's lime, also known as Type S or hydrated lime. You get it where masonry products are sold; it is different from non-hydrated lime used in agriculture, which will not work as a mortar admixture. The builder's lime will calcify (harden) over time, but its main purpose is to make the mortar more plastic and easier to use right out of the wheelbarrow or mixer.

Jaki and I mix in a wheelbarrow, and we've been doing this for over a quarter century. We add the ingredients to the barrow by the shovelful, using the following cadence, which greatly reduces dry mixing time. For the Portland mix:

- 3 shovels sand – 1 shovel sawdust – 1 shovel lime – 1 shovel Portland cement
- 3 shovels sand – 1 shovel sawdust – 1 shovel lime – 1 shovel Portland cement
- 3 shovels sand – 1 shovel sawdust – 1 shovel lime

Note that the introduction of the constituent ingredients in this way places the Portland cement one-third and two-thirds of the way down the mix. As only two shovels of Portland are used, there is none in the third line. With the masonry mix, a good cadence for adding material is:

- 3 shovels sand – 1 shovel sawdust – 1 shovel masonry cement – 1 shovel lime
- 3 shovels sand – 1 shovel sawdust – 1 shovel masonry cement – 1 shovel lime
- 3 shovels sand – 1 shovel sawdust – 1 shovel masonry cement

The numbers in these mixes refer to equal parts by volume, so always use the same size of shovel and load it the same way each time — small, medium, or heaping — depending on

3.4: The author, wearing cloth-lined rubber gloves and using an ordinary garden hoe, demonstrates mortar mixing at the Midwest Renewable Energy Fair in Amherst, Wisconsin. Credit: Jim Rhodes

the size of batch you want. Keep a separate shovel of the same size for the soaked sawdust, and keep the wet shovel out of the dry cementitious materials, which would soon make a mess out of the shovel.

You will need strong, cloth-lined rubber gloves throughout the project, including during the mixing process. Cementitious products, wet or dry, will eat nasty little holes in your hands. "Cement holes" become painful and take forever to go away. It may take a day or two to get used to working with your bulky gloves, but do it.

In an industrial-strength wheelbarrow, dry mix the goods with an ordinary garden hoe until everything is a uniform color. Then make a little crater in the center and add water. How much? Well, that depends on how wet the sand and sawdust are. On one occasion, the sand was so wet that I didn't have to add any water at all! But this is rare. For the first batch of the day, go easy on the first splash of water: add only a quart or two (a liter or two). Mix it in thoroughly and test it. Until you get really used to the mortar, we suggest the "snowball test." Toss a snowball-sized glob of mortar three feet in the air — one meter in Canada — and catch it in your gloved hand. If it shatters or crumbles, it is too dry. If it goes "sploot!" like a fresh cow pie, it is too wet. If it holds its shape, doesn't crack, and is nice and plastic, it is just right. (Note to experienced masons: You want stone mortar, not brick or block mortar. You folks know what I mean.)

If the mortar is too dry, add more water, remix, and test again until it is right. If too wet, add more dry goods in the same proportions until it is right. You can leave out the wet sawdust if the mix is really soupy, or you'll never dry it up enough.

Insulation

A cordwood wall derives its exceptional thermal characteristics from the insulated space between the inner and outer mortar joints. If this space is not insulated, the house will be very difficult to heat when it is cold outside: "Many have tried and a few were frozen." Silly, but you get the idea.

There are several choices for insulation in this space. Jaki and I did our first three buildings with fiberglass. But it was nasty stuff to work with (gets in the eyes and lungs), has a high embodied energy in its manufacture, and if it mats down with moisture, it may or may not fluff back again. Vermiculite, perlite, and other loose fill insulation work quite well but can be costly. Shredded beadboard seems, at first, like a good way to recycle materials, but it is virtually impossible to direct the stuff into the cavity. Ever hear of static cling? The little beads stick to the mortar, your gloves, your clothes ... they go everywhere, it seems, but where you want them. And the slightest wind is a disaster. So, thanks to the advice of Cordwood Guru and long-time friend Jack Henstridge (author of Chapter 5), we changed to using sawdust as insulation about 1980.

Sawdust is cheap, makes use of a waste material, and has an insulation value of about R-3 per inch — about the same as fiberglass. And it's easy to pour into the space with soup cans or small buckets. To retard against vermin, we mix builder's lime into our sawdust at a ratio of 12 parts sawdust to 1 part lime. Also, if the wall does take on moisture from whatever source, the lime will set up with the sawdust in the wall and form a kind of rigid foam insulation instead of a loose fill, while still doing the same job.

3.5: Top: Mortar goes down first, then insulation, then cordwood.

3.6: Bottom: Subsequent courses follow the same mantra: mortar, then insulation, then wood.

Building a Cordwood Wall

Carry the mortar to the site in the wheelbarrow. You can work out of the barrow or load up a metal or plastic mortar pan for convenient access to the "mud."

The foundation should be swept and dampened slightly. Several sizes of prepared log-ends should be within arm's reach. For discussion, we'll assume a 12-inch-thick (30-centimeter-thick) wall. Picture the width of that wall's footprint divided into thirds, like a French or Mexican flag. One-third part mortar, one-third sawdust, one-third mortar. We pass out patterns called M-I-M sticks to our students to help them gauge this proportion. So Mortar, Insulation, and Mortar are graphically marked right on the stick, which can be a 12-inch piece of scrap board. By the second day, students are doing pretty well without consulting the M-I-M stick. Make two or three for your project.

The building mantra is: Mortar. Insulation. Wood. This is the order to take. If you get out of this order, there is much wasted time in an already labor-intensive process. We will often see a student trying a log-end in a space, for example, before the insulation is installed. "I was just seeing if it would fit," is the common excuse. We point out in our friendly way that if the insulation was already in place, and the log-end happened to fit, it could be left in place, instead of placed on the floor again while the sawdust is installed.

So, using your gloved hands, grab a glob of mud and plunk it down on the foundation, about an inch (two-and-a-half centimeters) thick. (If your M-I-M stick is made from one-inch thick material, it can double as a mortar depth checker!) Keep adding more mud, extending the 4-inch-wide (10-centimeter-wide) mortar bed for three or four feet (about a meter). Now do the same thing for the other parallel mortar bed.

Next, with a small spouted bucket, pour in the lime-treated sawdust insulation between the two mortar beds, up to the same level as the mortar.

Now, grab a log-end and set it on the mortar, spanning the insulation. A slight vibration motion back and forth is all that is needed to establish a suction bond to the mortar. (Later, this suction bond becomes a friction bond, which is the best you can hope for with cordwood. There is no chemical bond between wood and mortar.) The next log-end is placed next to the first, leaving about an inch between log-ends. Continue until all the mortar is covered.

You can continue laterally around the home, a course or two at a time. If you are working with cordwood panels within a post and beam frame, you can work three or four courses high at a time.

The mantra doesn't change on the second course. Put the mud down first, following the hills and valleys formed by the first course of wood. Then comes the sawdust. Use your gloved fingers and thumbs to pre-settle the sawdust in the spaces between log-ends. Bring the sawdust up to the level of the mud. Now, select a log-end that has the same shape as defined by the previous mortar course. If you keep a good variety of large and small log-ends nearby, this will become easy with experience. Again, place it with a gentle vibrating set. You don't have to pound it in, although sometimes a gentle tap with a small hammer is helpful. If the log-end doesn't seem to want to "sit," it is almost always because you've used too much sawdust, which is now trying to spring the log-end back up again. Remove a little sawdust and try again. The other possibility (rare) is that an irregularity on one or both of the log-ends is getting in the way.

Be sure to leave about an inch of space between log-ends, so that you can get in with your pointing knife. There's nothing more frustrating than trying to get a ¾-inch-wide pointing knife into a ½-inch space. Which leads us to pointing.

Pointing

Pointing, also known as "tuck-pointing" or "grouting," is, in our view, a very important part of the masonry process. With cordwood masonry, pointing accomplishes several purposes. First, good stiff pointing maximizes the friction bond between wood and mortar. Remember that there is no chemical bond between the two.

Second, pointing beautifies the wall. Jaki can point a poorly laid wall, complaining all the time, and make it look better than a well-laid wall that is not pointed or poorly pointed. Do both: lay it up well and point it well. I have seen the opposite: a non-pointed poorly laid wall, which gives cordwood masonry a bad name.

Third, good pointing smoothes the mortar, making a more water-repelling surface on the outside, and a less dusty interior.

Fourth, if the pointing is recessed slightly, say ¼-inch to ½-inch, and all the log-ends in the wall shrink, it will be easy to conduct a repair (see Chapter 12 for repair suggestions). Recessed pointing also looks better. The log-ends are the defining features of a cordwood wall. Having them sit proud of the mortar is what gives the pleasing texture of the surface, similar to stone masonry.

You'll need a few pointing knives. The tools made for brick and block raking are not suitable. They are designed for straight ⅜-inch mortar joints. Jaki and I get pointing knives from thrift stores and garage sales. We recently purchased eight beauties from an antique store going out of business. One dollar the lot. We look for non-serrated butter knives. I like the ones that are almost an inch wide, but it is good to have a variety, and we even keep one or two with narrow ends for use where log-ends were laid too close together. Bend the last inch of the knives to about a 15- or 20-degree angle, so that you can get the business end in close to the work without your knuckles getting in the way.

Jaki, queen of the pointers, advises doing a "rough pointing" to the wall first, using just the rubber gloves. Remove excess mortar and catch it in your gloved hand. Then use your knife to press it off your palm or fingers into any gaps. "Borrow from Peter to pay Paul," she says — not good economy, perhaps, but it works with cordwood pointing.

For the finished pointing, press quite stiffly with the knife blade, and draw the mortar out smooth, trying to remove knife marks, if possible. How fussy you want to be is up to you, but keep a consistency of style. Do not over-point. You can be so fussy, going over and over the work, that you will simply bring a lot of water to the surface, which will cause cracking of the mortar within a few days. Been there, done that.

3.7: The mortar joint is simultaneously tightened, recessed, and smoothed by applying pressure with a pointing knife.

Well, that's basic Cordwood 101. Don't forget to wash your gloves and tools, and cover the work for the night.

There are additional construction tips and techniques in *Complete Book of Cordwood Masonry* (Sterling, 1992), as well as in the rest of this book.

Part Two
The State of the Art

4 • Stackwall Construction: The Double Wall Technique 37
Cliff Shockey

5 • The Lomax Corner 43
Jack Henstridge

6 • A Round Cordwood House with 16 Sides 49
Rob Roy

7 • Octagons, Hexagons, and Other Shapes 57
Rob Roy

8 • Bottle Designs in a Cordwood Wall 63
Valerie Davidson

9 • Patterned Cordwood Masonry 71
Rob Roy

10 • Electrical Wiring in Cordwood Masonry Buildings 79
Paul Mikalauskas and Mike Abel

11 • Using Cement Retarder with Cordwood Masonry 85
Rob Roy

12 • When It Shrinks, Stuff It! 93
Geoff Huggins

13 • A Mobile Home Converted to Cordwood 99
Al Fritsch and Jack Kieffer

14 • A Shop Teacher's Approach 103
James S. Juczak

15 • Paper-Enhanced Mortar 109
Alan Stankevitz

CHAPTER 4

Stackwall Construction: The Double Wall Technique

Cliff Shockey

(Editor's Note: Don't worry! Stackwall construction is just what Cliff and lots of other Canadians call cordwood masonry. The chapter title is also the same as the name of Cliff's book on the subject [see the Bibliography]. For the sake of uniformity and clarity, we'll use "cordwood masonry" here, for the most part. The term "stackwall" has most commonly been retained in the literature to describe the built-up corners made of dimensional wooden blocks. Thus: "stackwall corners." [Chapter 5, by another Canadian cordwood mason, tells of an excellent system for making such corners.])

Introduction

IN 1976–77, I TOOK A COURSE IN CONVENTIONAL LOG BUILDING through the local community college, but after the course, I wasn't convinced that horizontal log construction would be practical for our severe prairie winters. I was also a member of the Solar Energy Society of Saskatchewan, where I learned a lot about energy efficiency.

One day, while thumbing through an issue of *The Mother Earth News*, I came across a picture of a cordwood house built by Jack Henstridge of New Brunswick. "I can do that!" I said. The wheels immediately began turning, and soon I came up with the concept of double stackwall (hereinafter "cordwood masonry" or "cordwood") construction, and knew that I should combine the idea with sound solar design principles. With our cold winters, anyone building a home should design it to be as warm and as comfortable as possible. The double cordwood technique has worked extremely well for us and ensures security, warmth, and comfort in extreme climatic conditions.

Solar Design

Energy efficiency in construction requires a little common sense at the design stage. For example, the sun is a great source of energy, so why not take advantage of it? Design and build to let the sun's rays help heat your home. Image 4.1 shows how I used a 4-foot (1.2-meter) overhang on the south side of my buildings to take advantage of the sun's energy. When the winter sun is low, it floods the large south-facing windows for passive solar gain. As the sun rises higher in the spring, less heat enters the house because of the extended overhang. Therefore, the house remains cooler during the warm months.

We eliminated windows on the north side of the house because there is no solar gain from that direction. Also, heat loss through north-facing windows would be heavy because of the prevailing north winds.

Foundations and Under-floor Radiant Heat

The foundation has to be broad enough to support a wide cordwood wall. Our double wall technique has 24-inch-wide (60-centimeter-wide) walls, so the foundation choice is particularly important, and will probably mean spending a fair amount of money on concrete. The best method of doing this is with a thickened edge floating slab.

If the concrete floor is poured independently of the footings, you can also incorporate a radiant heating system in the concrete slab, as we did in 1985 in our stackwall insurance office building in Vanscoy, Saskatchewan, and again in our house addition built in 1990. Rubber or plastic tubes are laid in the concrete in a kind of a labyrinthine pattern to circulate hot water through the floor. At the insurance office, the water is heated by a small natural gas-fired boiler and is circulated by an electric pump. We are very pleased with this system, as it has proven to be a very comfortable and efficient way of heating. In February of 1999, I checked to see how much it costs to heat this 814-square-foot (76-square-meter) office building.

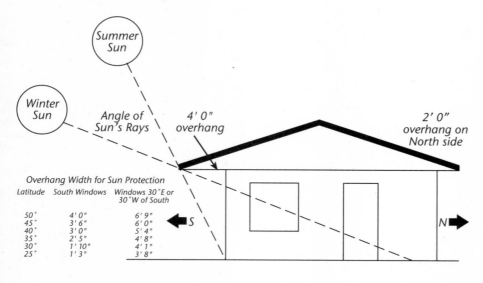

4.1: Passive solar design.
Credit: Rob Pichelman

The natural gas bill for the month of November 1998 was Can$32.90 and for January 1999, it was Can$40, so you can see that the building is very energy efficient. Keep in mind that we are in an extremely cold part of North America.

Under-floor radiant heat should be professionally installed. The installers will insist upon 2 inches (5 centimeters) (R-10) of extruded polystyrene insulation under the slab, which is typically 4 to 6 inches (10 to 15 centimeters) thick. A thermal break of insulation is also placed between the slab and the footings. The floor can be painted, stained, or covered with tiles or slate.

Incidentally, cordwood masonry is a good choice for commercial buildings. The pleasing esthetics of the building has a positive impact on the people who work and visit there. The building is comfortable in terms of its intangible "atmosphere" as well as in its thermal characteristics. With buildings such as stores and restaurants, the curiosity value of cordwood masonry could actually serve as a drawing card to attract new customers.

The Double Wall Cordwood Technique

The double wall technique involves building an 8-inch-wide (20-centimeter-wide) outer cordwood wall and another inner cordwood wall of the same width. The space between the cordwood walls is also 8 inches, and is occupied by: some inexpensive sheathing on the inside of the outer wall; 8-inch fiberglass batts; and a tight polyethylene vapor barrier just behind the inner wall. Image 4.2 shows a cross-section of the layers of a 24-inch-wide (60-centimeter-wide) cordwood wall built using the double wall technique.

On my first 600-square-foot (56-square-meter) cordwood house, built in 1977, I used built-up corners for the outer wall and laid up the inner wall within a post and beam framework. On my larger 1300-square-foot (121-square-meter) second house, I used the same method. In 1985, I decided on post and beam framing for both the inner and the outer walls of the insurance office, and did the same thing in a 392-square-foot (36-square-meter) addition to our second house in 1990. I now recommend the newer method, because it is faster and easier and enables you to get the roof on first and then work under cover. I used 8-by-8-inch timbers for my framing, but you can adapt your framing design to take advantage of material that you have salvaged or purchased at a good price. Image 4.3 shows a simple post and beam frame as it would be built for the double wall technique.

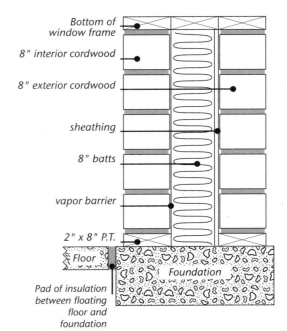

4.2: Cross-section of Cliff Shockey's double wall technique.
Credit: Rob Pichelman

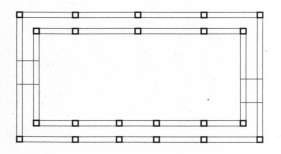

4.3: Plan view of simple post and beam frame for the double wall technique.

If you really like the appearance of the stackwall or built-up corners on the outside — they *are* very attractive — you could probably support a roof truss system by the inner post and beam frame, get the roof on, and build the outer wall afterwards with built-up corners.

With the post and beam method, the exterior cordwood masonry should be laid up first (using methods described in Chapter 3). Always work with heavy cloth-lined rubber gloves and leave between three-quarters and one inch of mortar between log-ends. The mortar mix I used is slightly different from Rob's and has worked well for us. It is, simply, 3 parts sand and 1 part masonry cement. On the exterior wall only, we add 1 part of screened and soaked sawdust. The sawdust slows the set on the outer wall, but the inner wall does not set too quickly if the outer wall is built first.

Another detail I do a little differently from Rob is to fasten a two-by-eight pressure-treated plank to the foundation as the base of my cordwood wall. This keeps the first course of the wood a little further off the concrete foundation, and on the inside, provides a place to fasten the vapor barrier. These plates can be fastened with anchor pins or concrete nails. My cordwood for all our buildings was cut from untreated cedar utility poles that I obtained for removing them along 6 miles (9.6 kilometers) of road. They were fairly regular of size, but I split some that had excessively large checks. These pieces were handy whenever a full round log-end wouldn't fit, such as at the end of a course, where the masonry meets a post or door frame. With a little imagination, you can make some very attractive patterns.

When the rough door and window framing is in, and the outside wall is complete (with all the additional features in place such as glass bottles, wagon wheels, dryer vents, etc.), I like to put 5/16-inch particle board (or any inexpensive sheathing) on the inside of the exterior wall. This acts as a backing for the insulation batts to come. Also, if you happen to be building higher than 8 feet (2.4 meters), the sheathing helps to stabilize the wall. It also serves as a barrier to help keep bugs or mice out of the insulation cavity.

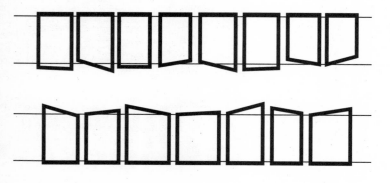

4.4: Plan view of double wall. Log-ends do not have to be cut perfectly in order to keep the interior and exterior surfaces straight.

The next step is installing the insulation. I like to use 8-inch (20-centimeter) (R-28) batts because they are fairly rigid. I find that they will stand on end without sagging down the wall. If you decide on thinner batts, you can pound nails part way into the wall and then, for stability, push the batts over them.

Next comes the vapor barrier. I feel that a tight polyethylene air vapor barrier is important for making an airtight, draft-free home. The vapor barrier is fastened to the pressure-treated two-by-eight base plate already mentioned, and also to the inner post and beam frame, the top plate, and all window and door frames. All seams in the vapor barrier must be sealed together with acoustical sealant over solid backing. With the vapor barrier in place, it is like living inside a big airtight bag, with openings only for things like doors, windows, plumbing, vent pipes, chimneys, etc. It is important to seal around all openings in the vapor barrier. Make sure that every seam is sealed before the walls and ceilings are finished, as it is impossible to get at it later. Some people have suggested that a very tight house necessitates the installation of an air-to-air heat exchanger to prevent stagnant air and promote sufficient air changes. I must say that we do not use an air-to-air heat exchanger and have not observed any problems with air quality.

After installing the vapor barrier, build the inner wall. Remember that the sawdust admixture is optional on the inner wall, providing that you are building within the plastic tent of the vapor barrier, which itself helps to retard the mortar set. Eliminating the sawdust results in a smoother finish to the mortar. You can do fine recessed finish pointing like Jaki Roy, or you can do a rough pointing with your rubber gloves like I did. Later, we cleaned the log-ends of loose mortar with an electric rotating wire brush, and then sprayed the wall with a spirit-diluted mixture of polyurethane. This brings the color out of the cedar in a very attractive way, and provides a surface that is a little easier to clean. The choice is yours.

One advantage of the double wall system is that any irregularities in log-end length or straightness of cut can be hidden out of sight toward the center of the wall (see Image 4.4).

People have asked me if the double wall technique isn't twice as much work as regular cordwood masonry. It isn't. Actually, you will need only two-thirds as much wood as with a standard 24-inch-thick (60-centimeter-thick) cordwood masonry wall. You'll mix only slightly more mortar. And you will have exactly the same amount of pointing.

4.5: Cliff's addition, shown under construction, was built in 1990. Credit: Cliff Shockey.

4.6: Cliff's main house, where he now lives, built 1978–80. Credit: Cliff Shockey.

On the upside, you have a highly energy-efficient structure, providing you use ceiling or roof insulation in scale with the wall insulation. As the double wall technique yields an insulation value of approximately R-40, we used R-56 insulation in our roofs.

Before you begin any building project, gather information from several different sources. In this way, you will be more likely to make well-informed decisions.

Above all else, take time to enjoy your project. Designing and building your own home can be one of the most satisfying endeavors you will ever experience.

CHAPTER 5

The Lomax Corner

Jack Henstridge

YOU HAVE TWO BASIC METHODS to choose between when it comes to square corners: post and beam construction with cordwood infilling or built-up corners. The post and beam method is well covered in the literature by such books as *Complete Book of Cordwood Masonry Housebuilding* (Sterling, 1992) by Rob Roy; *Cordwood Construction: A Log End View* (Self-published, 2002) by Richard Flatau; discussions in this current volume; and by many books devoted to timber framing itself. Built-up corners are discussed in *Stackwall: How to Build It*, 2nd ed. (A and K Technical Services, 1995) and in Rob's *Complete Book of Cordwood Masonry Housebuilding*. Check out the Bibliography.

Back in 1981, Bev London decided to build a cordwood house — not just an ordinary house but one that was a radical departure from anything that had been done previously. He had lots of wood and lots of time, but was short on cash, which is why he decided that cordwood was the way to go.

Bev's 24-by-48-foot (7.3-by-14.6-meter) house has a full basement and was completed for a total cash outlay of Can$36,500. If you count the basement in the total square footage (and you should because his equipment repair workshop, etc., are in there; it's all usable space) you come up with a cost of Can$15.91 per square foot! The secret of how he achieved this low figure was not simply the type of wall construction; it was also his choice of basement. Below grade he used concrete blocks, above grade, he used two-by-eights on 12-inch (30-centimeter) centers, which he figured would be more than adequate to support the weight of a single-story cordwood home.

5.1. Bev London's house in Coytown, New Brunswick.

5.2: Roy Telford's workshop in Upper Gagetown, New Brunswick.

Time has proved him correct — but as far as I know it is still the only cordwood building supported by this type of basement wall (see Image 5.1).

But what has all this got to do with square corners? you ask. Ah! Your patience is about to be rewarded. This unique structure was also the first to be built using the "Lomax Corner."

Once the basement was completed and the subfloor laid, it was time to lay the cordwood. I must admit that I was a little apprehensive, but Bev — who might have had a few doubts, too, mainly because of the number of people standing around shaking their heads — said, "Okay, boys, it's now or never. Let's get at it!" And the mixer started to churn.

One fellow had no doubts at all: Gary Lomax, an engineering type from Upper Canada who was familiar with this type of basement. Bev and Gary were buddies from way back and Gary had taken his vacation to come down and lend a hand. Gary was one of those individuals who had the ability to look at a sticky situation and figure a better way around it.

There's no question that built-up corners made by the traditional methods — laying individual wood corner blocks or "quoins" — is quite time consuming. I noticed Gary studying what we were doing. Before long, he had taken some of the squared corner pieces and hammered away at them mysteriously. Then he came over and said, "Here, try this!" And so the Lomax Corner was born.

Since that day, I have been involved in the construction of quite a few other stackwall-cornered structures ... and I wouldn't use anything else. In 2001, I helped build a home in New Brunswick using this method. The Lomax Corner not only speeds construction, it also makes for a much more energy-efficient junction. It allows the insulated space within the wall to continue right around the corner, whereas in previous corners, there was always a solid section of mortar that acted as a "heat wick." With the Lomax method, the two mortar walls are totally isolated from each other by the insulating material.

Let's examine the construction, step by step. For the purposes of this example, we'll assume 16-inch-thick (40-centimeter-thick) walls and corner units made from full-sized, rough-cut 4-by-4-inch timbers. But keep in mind that the same technique can be adapted to other available dimensions, such as 4-by-6-inch or 6-by-6-inch. Whatever you use, uniformity is important.

Step 1: Make a materials list. Four-by-four material is usually purchased in 8-foot lengths. It's quite easy to figure how many lengths you will need. Simply take the height you plan to build your wall — in inches — and divide that by 6 inches. For example: With an 8-foot (96-inch) wall, divide 96 inches by 6 inches, which gives 16 pieces. So there are 16 corner pieces in each corner. Why do we divide by 6 inches when using four-by-fours? Each corner unit is made with two 1-inch tie pieces (full-sized 1-by-1-inch stock works well) on each side. Four inches plus 1 inch plus 1 inch equals 6 inches. The corner unit is made with two lengths of four-by-four. Let's assume that we're going to make them 2 feet long. In a 16-inch thick wall, this provides an 8-inch tie-in to the ordinary cordwood masonry of the sidewalls.

Now, 16 corner blocks times 2 pieces each, times 2 feet long, equals 64 linear feet of 4-by-4-inch material (16 × 2 × 2 feet = 64 feet). So, with 8-footers, each corner will take 8 lengths. Multiply this by the number of corners in your design. Four corners, for example, require 32 lengths of 8-foot-long, 4-by-4-inch material. Always purchase a couple of extra lengths. Eastern white cedar, in some cases, may have a hollow center when you cut into it. This material isn't cheap, so select carefully. You'll need a few extra pieces, anyway, for some short filler blocks later on.

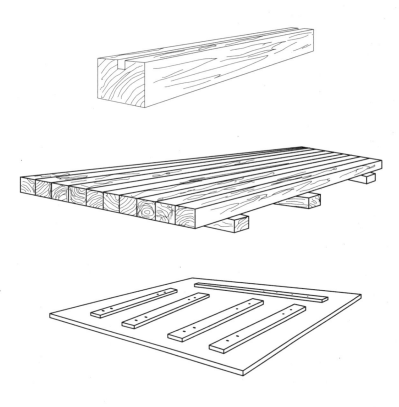

5.3, 5.4 and 5.5

You'll also need plenty of rough-cut, full-sized 1-by-1-inch material for the tie pieces. In the example used here, each Lomax Corner unit uses four 12-inch tie pieces, so you will need 4 linear feet for each corner unit, or 64 linear feet (16 pieces × 4 feet) for each corner. You can make your own tie pieces quite cheaply with a table saw by ripping any one-inch rough-cut stock into full 1-inch-wide pieces. Set the table's fence to give you full-sized one-by-ones, to maintain uniformity throughout the job. Make plenty. Knots in a one-by-one render a piece useless, except as kindling.

Step 2: Get in a groove. Lay all your pieces out and rip a single groove down one side with your chainsaw. To speed up this chore, rank the pieces so that you can groove a bunch at a time. The groove greatly improves the friction bond between the corner units and the mortar. Remember that there is virtually no chemical bond between wood and mortar, so

5.6: *These two units are ready for installation into a Lomax-type stackwall corner.*

5.7: *This overhead view shows how Lomax units are installed relative to each other. Notice grooves to key the unit into the mortar. The space in the middle is filled with sawdust insulation, separating the inner and outer mortar joints.*

5.8: *The 1-inch-thick tie pieces maintain the 2-inch space between the corner units. Rose, the dog, keeps her eye on the work. Credit (all three images): Jack Henstridge*

we want to maximize the friction bond. You can even groove both the top and bottom surfaces of the unit, as some have done in recent years.

Step 3: Cut your pieces to length. In our example, we'll need to crosscut each 8-foot length of 4-by-4 into four 2-foot pieces. To speed this up, mark the pieces with a lumber crayon while they are ranked, as in Image 5.4.

Step 4: Take the time to do a jig. No, you won't need a fiddler. I'm talking about an assembly jig. This can be built several ways. Master Mortar Stuffer Roy Telford came up with the idea of using an old piece of ¾-inch plywood for the base of his jig. This way, you can set up your assembly almost anywhere (Think shade tree!).

The four parallel pieces (1-by-3-inch strapping works fine) are set on the board to hold the two 4-by-4-by-24-inch corner pieces so that the pieces are 16 inches from one outside edge to the other (or whatever thickness of wall you are building). The crosspiece at the top of the drawing is the stop-block. This jig holds everything square and parallel for nailing on the tie pieces (see image 5.5).

Step 5: Assembly. Place the four-by-fours in the jig and nail the first tie piece 2½ inches from the end next to the stop-block and the second one at 11½ inches in. This assumes the use of one-by-one tie pieces. Make sure these tie pieces are accurately placed — you should mark the assembly board with a lumber crayon. They must be accurate so that the Lomax units will stack up properly when placed in position at the corner.

When the two top tie pieces are secured, remove the two blocks from the jig, turn the unit over, and nail two more pieces on the bottom side in the same position. Your corner unit is almost complete.

You can make short blocks — 6-inch lengths are fine — from extra or leftover four-by-four material. These 6-inch filler blocks could be positioned in the jig when you are assembling, or placed in now. If placed in now, be sure that this filler block is even with the two side pieces (this can be seen clearly in the photos). These blocks prevent a very large unattractive expanse of mortar at the outside end of each unit, but they also help to retain the insulation during construction.

Your corner block, or Lomax unit, is now ready to go into the wall. However, it is advisable to build up a bunch of them, as it is easier to nail (temporarily) four of the units together, alternating them crisscross fashion as they would be when installed in the wall. For this, you can use some scrap strapping to hold them in position.

Now all you have to do is stuff the mortar in the gaps and pour your insulating material into the space in the middle as you build the cordwood walls to it.

Study the photographs carefully, and I'm sure you'll get the picture (no pun intended) of how it all goes together. It may sound like a lot of "fiddly" work, but believe me when I tell you that it is the easiest, most accurate way to construct built-up corners. Thank you, Gary Lomax!

5.9: In this large Queenstown, New Brunswick home, completed in 2001, longer Lomax units help break the vertical joint of the shorter units. Note the use of various bracing to hold window frames and Lomax units plumb until the mortar sets. Credit: Jack Henstridge

(Editor's Note: Author "Cordwood" Jack Henstridge is one of the founding fathers of the modern cordwood movement. In the introduction to the original version of this article, (CoCoCo/99 Collected Papers, pages 15-21), Jack said, "Anyone who knows me, has read my articles, or has listened to my presentations, rapidly realizes that I'm not exactly what you'd call a Big Corner Fan." Jack's favorite cordwood style is the curved wall technique, and he has influenced a large number of people — including Jaki and I, to favor that method. But I share with Jack an appreciation for all the beautiful and successful stackwall-cornered homes that have been built throughout North America, and particularly in Canada. I agree with him that it makes little sense to construct built-up corners by any other but the Lomax method. The corners are faster and easier to build, more regular, and almost certainly stronger.

Since publication of the earlier version of Jack's article, many others have built successful rectilinear homes with Lomax Corners. And some, including Ed Cote of Childwold, New York, have built half of the corner units six inches longer than the other half. When they are placed in the wall, crisscross fashion, Ed makes sure that on each side of the corner, he alternates a 24-inch-long unit with one of, say, 30 inches. This eliminates the long straight vertical line-up of units, which could potentially cause a shear crack in the mortar. Staggering the units ties the corners better into the main run of the cordwood wall. When I mentioned this to Jack in a phone conversation in January 2002, he said that he'd done something similar at the large home he'd recently helped build in 2001 in Queenstown, New Brunswick, alternating two courses of 24-inch Lomax units with one of 36-inch units.)

CHAPTER 6

A Round Cordwood House with 16 Sides

Rob Roy

THIS CHAPTER MUST BEGIN with the acknowledgment that it was Bunny and Bear Fraser who did all the hard work at their home, Hutchnden House, in Coe Hill, Ontario. I am just the messenger. Others have built 16-sided cordwood buildings before the Frasers, but Bunny and Bear's system is a great improvement in minimizing material requirements, thanks to their clever use of the "temporary post." Also, Bear kept a written record of their project. Some of his notes, gathered in a sidebar, give a bit of the flavor of the project and share some useful building tips.

Some readers may have a problem with calling a 16-sided building a "round house." Okay, okay, 16-sided is *not* absolutely round, but it's darned close to it, very much more so than an octagon, for example, a popular shape for cordwood houses (see Chapter 7 for a discussion of other polygon shapes). In his entertaining novel *Flatland* (Penguin, 1998), E.A. Abbott tells of a world of two dimensions, where the inhabitants are regular geometric figures (triangles, squares, hexagons, octagons, etc.) who move about by sliding around their tabletop world. In Flatland, the more sides one has, the higher that individual's standing in the social hierarchy. If someone has 32 sides, they are given the title of "honorary circle," because further differentiation is meaningless. I submit that 16-sided is even better than 32-sided: the home looks round and feels round, and yet the design has lots of flat sections of perimeter wall, a godsend when dealing with things like kitchen cabinets and the placement of certain appliances and furniture.

In short, Hutchnden House is honorarily round. It *looks* round — see the pictures — and it feels round. Jaki and I know this because we've stayed several nights at the home, which Bunny and Bear used to operate as a bed & breakfast. Oh, and for those slow on the uptake, Hutchnden refers to a "bunny hutch" and a "bear's den." And, by the way, the Frasers sold their home to another couple who still operate it as a B & B. (For information, go to: www3.sympatico.ca/vanden, or call Annette and Al Vandendriessche at: (613) 337•5177).

Before getting into the *how*, a word must be said about the *why* of the 16-sided frame. The advantage of cordwood infilling within a post and beam frame — over either the stackwall (built-up) corners system or the load-bearing curved wall shapes — is that the roof can be put on first, affording the builders the luxury of building the cordwood walls under cover. The roof acts as a giant umbrella and provides protection from both rain and direct sun. The Fraser Frame system introduces this compelling advantage to the round designs. Also, as many have found, code officials love to see that post and beam frame in the design. Despite the hundreds of load-bearing cordwood homes all over North America, some code enforcement officers are reluctant to sign off on what they perceive to be a glorified stack of firewood.

The Fraser Frame

The Fraser Frame method uses a minimum of framing timbers: just 32 posts, and 32 girders that join the tops of the posts. All members can be cut from 8-foot timbers with very little waste. Bear chose to use pressure-treated 6-by-6-inch timbers for his frame (the green kind). They are strong, rot resistant, attractive, and readily available. The frame need not be exposed to any part of the interior, thus avoiding the off-gassing of poisons into the living space. (Be sure to use eye and nose protection when cutting, even outdoors.) Other types of 6-by-6-inch stock can be used, and the method could be adapted to other diameters of buildings. The Hutchnden House walls are 18 inches (46 centimeters) thick. With a smaller building, with 12-inch-thick (30-centimeter-thick) walls, four-by-fours would be in scale. With a larger diameter building, 24-inch-thick (60-centimeter-thick) walls might be appropriate, framed by eight-by-eights.

The best way to show how to do the Fraser Frame is by "narrating" several photographs, provided by Bunny and Bear themselves. A picture, after all, is worth 1,000 words, so sit back and pretend that you're viewing these pictures as slides at a workshop, just as our students do.

We begin with a look at the entire completed framework, consisting of 16 posts on the first floor, joined by 16 girders. After the first-story frame is completed, the radial floor joist system for the second story is installed. These joists consist of doubled 2-by-

6.1: Credit: David Fraser

12-inch material, about 2 feet on-center (24 inches [60 centimeters] on-center) where they pass over the girders. (You can peek ahead at Image 6.2, if you like, to see this clearly.) While we have 6.1 on the screen, note that the frame divides the upstairs and the downstairs into 16 panels each — 32 in all — each measuring roughly 8 feet square (2.4 meters square). This is great for you goal-oriented people who like to see a little mathematical progress to your work, as in: "Well, we've got eight panels done. Our cordwood is 25 percent completed!"

Note, also, the temporary diagonals which give rigidity to the framework. They stop the building from "racking," even during heavy winds. Later, of course, the cordwood masonry does the same job. The roof, with its 36-inch (90-centimeter) overhang, provides a substantial umbrella over the entire site.

6.2: Credit: David Fraser

Image 6.2 is a good view of the post and beam frame, the temporary diagonals, and the radial floor joist system for the second story. Note that a doubled 2-by-12 falls directly over each post, and three more fall along the girder between posts. At this point, then, the floor joists are on roughly 2-foot centers (24 inches [60 centimeters] on-center). Due to the nature of a radial support system, as joists head for the center of the building they get closer together; on the deck overhang, they get a little further apart but remain extremely strong.

6.3: Credit: David Fraser

There are two other points to be made about this picture. First, at the center of the home, you can see an octagonal framework that defines a 10-foot (3-meter) diameter room. The first floor joists cantilever upon this frame and meet at the center of the home, without benefit of a central post. An identical room on the second story performs the same function for the truss system that supports the roof. The second point to look for here is that a couple of 2-by-6-inch pieces are scabbed to the doubled floor joists just at the point where they pass over the first-floor posts. These short pieces effectively broaden the joists at this point to a full 6-by-6-foot area, thus providing full bearing for the upstairs ring of posts. So, the roof

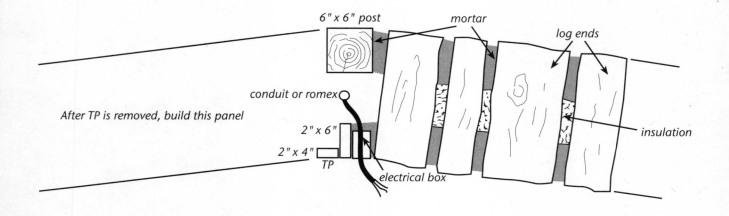

6.4: Credit: Rob Roy

thrust is passed down through the upper story posts, through the floor joists and their stabilizing scab pieces, onto the first-story posts and, finally, to the concrete foundation. This is a very tidy stabilizing detail which you might not think of, but Bear's camera captures it nicely.

In Image 6.3, for the first time, we see Bear's clever innovation that makes the system possible with cordwood masonry — something I call the "temporary post," or TP for short. Bear used four or five of these during construction, and makes them from a two-by-four nailed to a two-by-six for stiffness. The TPs are set up as shown here and in Image 6.4.

In a truly round house, like Earthwood, each log-end aims toward the center of the building. But with a 16-sided building whose individual panels are about 8 feet (2.4 meters) wide, only the log-ends toward the middle of the panel aim for the building's center. The ones near the posts actually point to a spot about 4 feet (1.2 meters) off-center. If a second permanent post were to be installed where the TP is, we'd have two new problems: First, we'd be using twice as much material as is necessary, and second, a permanent post at this position would make the panel width quite a bit less on the inside part of the wall than on the outside. The inner mortar joints would have to be very much smaller than their counterparts on the outside of the wall. In short, the log-ends won't fit in well. The use of the temporary post avoids both problems.

Bear sets up, say, four consecutive TPs along the inner surface of the cordwood wall, as per Images 6.3 and 6.4, being careful which way the two-by-six component faces. The flat edge of the two-by-six falls along a line from the edge of the permanent post to a point roughly 4 feet (1.2 meters) off-center. This flat edge of the smoothly finished two-by-six is lightly oiled to facilitate its removal later on. Bear sets up the TPs so that he can build every second panel. If the panels are numbered 1 through 16, for example, he will first lay up cordwood in panels 1, 3, and 5. After those panels are completed, he removes the temporary posts, leaving the regular 1-inch mortar joint behind, as you can see if you look forward to Images 6.5 and 6.6. Later, the alternate panels (2, 4, etc.) can be infilled with cordwood masonry, laying the fresh mud right up to the 1-inch mortar joint that was left behind. If ten days or so have transpired since the previous work was laid up, you may wish to paint the 6-inch-wide (15-centimeter-wide) edge of the mortar joint with some Acryl-60 (Thoro Corporation) or an equivalent mortar bonding agent.

When you lay up the panels against the temporary posts, don't worry about the sawdust insulation at the extremities of the panel. This insulation is easy to install when you do the other panels, the "even" panels, after the TPs are removed.

Image 6.4 is another view of 6.3, showing a little more clearly how the temporary post position relates to the permanent post. Also, I have drawn in an electrical box, laid up against the TP. If it is for a duplex receptacle, the box would be about 16 inches (40 centimeters) off of the floor. If it is for a light switch, it would be installed about 48 inches (120 centimeters) off of the floor. The Romex conductor or conduit comes out the back of the box and runs through the insulated space. (See also Images 6.5 and 6.6.

In Image 6.5 we see the top of a cordwood panel after the TP is removed. An internal plate beam made of 6-by-6-inch timbers is screw-nailed to the underside of the floor joists and the cordwood panel is built up to it. The space between the inner plate beam and the 6-by-6-inch pressure-treated girder on the outside makes a great chaseway for electrical installations.

Image 6.6 shows the position of the first log-end of the new cordwood panel about to be built. Remember, it aims not for the center of the home, but

6.5 left, Credit: David Fraser.

6.6 right, Credit: David Fraser.

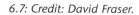

6.7: Credit: David Fraser.

for a point about 4 feet (1.2 meters) off-center. Note the installation of the electrical box and wiring in each picture. (See also Chapter 10 for more electrical installation ideas with cordwood masonry.)

Image 6.7 is a view of the completed Hutchnden House. Note that the panels in this picture — of the western hemisphere — all contain windows or sliding glass door units, usually centered in the panel. Some of the unseen panels, hidden on the other facets of the building, are made up entirely of cordwood masonry.

Can the Fraser Frame Be Used Under an Earth Roof?

The answer is, Yes. In fact, builders in different parts of the country are doing this now. Jim Juczak's 18-sided home (see Chapter 14) is engineered for the load, although the earth was not in place as of April 2002. The order of events is as follows:

1. Start with the Earthwood round house plans, including the octagonal or similar inner frame which shortens floor joist and rafter spans. Include a central column as per Earthwood, Pompanuck, or the Juczak home.
2. Incorporate a Fraser external frame into the plan. At Earthwood, this would involve a 6-by-6-inch post at every second primary rafter around the perimeter. Adjacent posts are connected by 6-by-6-inch girders.
3. Erect all framing and the central column. Include floor joists, floor or subfloor, roof rafters, and roof decking. Brace the entire frame with plenty of temporary diagonal boards, screwed onto the frame for easy removal.
4. Install the waterproofing membrane and cover it, if necessary, to protect it from the sun's ultraviolet (UV) rays. The W.R. Grace Bituthene™ 4000 membrane, for example, cannot be left exposed to full sunlight for very long. Do not put the crushed stone drainage layer or the earth on at this point, as the 6-by-6-inch girders do not have sufficient bending strength to handle a wet earth roof.
5. Infill all the panels with cordwood masonry. If there is a window or door in the panel, make sure that there is a heavy enough lintel over the window or door frame to carry the load.
6. After the last panel is infilled, wait a week for the mortar to gain strength, and then install the various layers of the earth roof. See my *Complete Book of Cordwood Masonry Housebuilding* (Sterling, 1992) or *The Complete Book of Underground Houses* (Sterling, 1994) for a complete discussion of the earth roof layers.

The Bear Wisdom

David (Bear) Fraser

On the cleared and leveled site, we surveyed and staked what by this time was a more or less, almost definite, possibly final floor plan on the ground, and marked out the room sizes with fluorescent tape.

Because we wanted to do the job right the first time, we decided on the extra security of gluing and screwing rather than just nailing the flooring into place. It took more time and money, but we were happy knowing that the only creaking our home would hear would be from human joints.

We bought a used mortar mixer, having decided that the only way we were going to mix more than 20 cubic yards (15.3 cubic meters) of mud was with mechanical assistance. We knew this might look like treason to some of our forefather-oriented friends. At Earthwood, the Roys showed us that mixing mortar by hand in a wheelbarrow was ethically wholesome, morally worthwhile, and spiritually uplifting. But our budget would not cover wages for more than one individual whose sole job was to keep up to three mortar stuffing teams supplied with mud, sawdust, and cordwood.

On July 11, 1995 the first of 32 panels of logs was completed. The panel took five days, which scared the hell out of Bear. At that rate, we'd still be laying up logs at the end of February 1996!

As time passed, we were able to cut our cordwood masonry rate by more than half; a team of mortar stuffers could average $2\text{-}1/2$ panels in a five-day workweek. Things were starting to take on a tentative air of optimism.

One day we discovered that we'd used up our supply of the dry sawdust we mixed with lime for our insulation, so we had little choice but to start using the pile that had been left uncovered and was, in fact, quite moist. I packed some of the moist sawdust and lime mixture in a box and promptly mislaid it. When I found it weeks later, I noticed it was quite solid and had taken on characteristics not unlike rigid-foam insulation! A glorious accident had become another of Bunny and Bear's tricks of the mortar-stuffing trade. And using the moist sawdust had other benefits: you don't have to wear a mask or worry about breathing in noxious lime dust; and the mix will not move about once it sets, so air pockets will not develop due to settling.

On October 4, 1995, Jack and Blaise laid the last log, drawing the masonic chapter of our house to a close. They'd done a super job and we appreciated their help for all those weeks. It's hard to believe that the cordwood laying and mortar stuffing had been completed in less than three months.

On November 10, we began the Herculean task of moving in and unpacking. By the 30th, the last contractor had left and we were settling very enthusiastically into a wonderful home, built with love.

Bunny and Bear, incidentally, decided on a more conventional shingle roof. Their roof has 16 facets, each having a pitch of 6 in 12 (6:12). Even adding the local snow load, their maximum roof load is much less than the Earthwood roof, which is engineered for 165 pounds per square foot. Because of the lesser load, they are able to go with very much greater spans with their truss system than we could at Earthwood, even with our heavy 5-by-10-inch rafter system. Half of their trusses go all the way to the center of the house, where they are sort of ganged together. The other half, which I call the secondary trusses, stop at the 10-foot (3-meter) diameter room.

CHAPTER 7

Octagons, Hexagons, and Other Shapes

Rob Roy

Let me say up front that I have never built either an octagonal house or a hexagon. I have built several octagonal post and beam frames, however, as well as eight-sided roofs. And I have done a lot of infilling of post and beam frames with cordwood masonry and so know some of the design considerations to watch out for. I've spoken with builders who have built cordwood octagons. And, at design sessions, I've spent a lot of time working out corner details with those that have eight- or six-sided houses in mind. Incidentally, one of our former students built a nine-sided cordwood home in New Hampshire, just for the challenge, I think. It came out fine, but it must have been a nightmare of complicated angles and detailed carpentry. It's best to stay clear of designs with really odd angles, unless you're already skilled in this sort of thing.

Octagons are popular. They appeal to people who want something out of the ordinary but for whom the move to a round house is just too much of a departure from what they're used to. Or, they think that to build a round house must be difficult, despite the fact that round was humankind's first design choice. Perhaps early *homo sapiens* took their cue from the animals around them. The problems with octagons (and other polygons), however, are making use of the unusual shapes of rooms — this is also true of round houses — and addressing the corner detail of the post and beam frame so that the cordwood masonry has something to bear against.

Creative Framing for Polygons

An octagon has a 135-degree angle where two sides come together. Sawmills do not cut timbers at a 135-degree angle or at other strange angles, such as the 112½-degree angles shown in Image 7.1. The builder, then, must come up with a way to make corner posts that approximate these angles, so that the cordwood wall has a nice flat post surface to bear

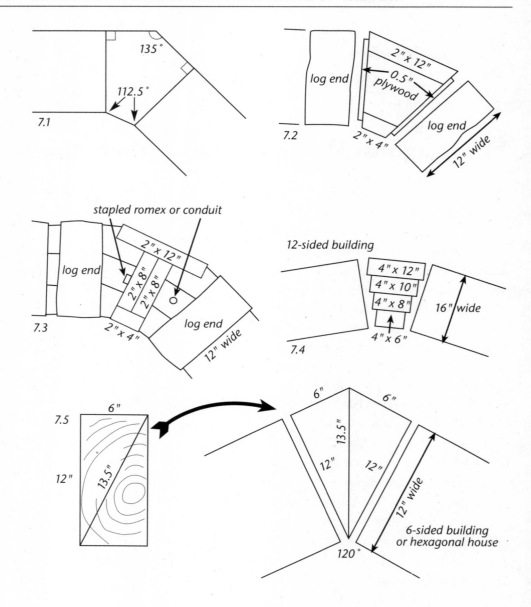

7.1, 7.2, 7.3, 7.4 and 7.5
Credit: Rob Roy.

against. The log-ends, in other words, need to be parallel to the edge of the post, just as in a square building. Here are some suggestions. The first three — which refer to Images 7.1, 7.2 and 7.3 respectively — work with octagons, but the ideas could be adapted for other polygons. The third suggestion works with a 12-sided building and refers to Image 7.4. The fourth idea is for a hexagonal building and refers to Image 7.5.

1. With an adz, hew a tree trunk by hand to the required shape (Image 7.1). An alternative to hand hewing is to set up an Alaskan chainsaw mill to make the required cuts. See below.
2. Build up a post of, for example, a 2-by-4 and a 2-by-12, scabbed together as required with pieces of ½-inch plywood (Image 7.2) or internal 2-by-8 pieces (Image 7.3). These methods make good posts for a 12-inch-thick (30-centimeter-thick) wall and provide space for Romex conductor or flexible conduit.
3. Build up the required post with ever-decreasing dimensional lumber, as per Image 7.4. Here a rough-cut 4-by-12 is used on the outside. Then a 4-by-10 is nailed to it, then a 4-by-8, and finally a 4-by-6. This makes a sturdy post for a 12-sided building with 16-inch-thick (40-centimeter-thick) walls.
4. A clever hexagon design is possible if you are able to rip a timber diagonally. The timber must be twice as broad as wide, as in a 6-by-12. Rip the piece corner to corner. Then align the diagonal cuts together to form a post for use with a hexagon, as per Image 7.5. A skilled chainsaw operator with a freshly sharpened chain can do this. A special ripping chain works much faster than a normal saw chain, which is made for crosscutting.

If it is your intent to make rough-cut post and beam timbers out of your own trees, you should seriously consider purchasing a chainsaw mill. There are several different manufacturers of chainsaw attachments to do the job, and prices vary from about $40 to $175. A ripping chain and a powerful (3.8-cubic-inch, 20-inch bar) saw is a must for use with attachments like the Alaskan Small Log Mill or other basic Alaskan Mills. A good source for all sorts of chainsaw equipment, including chainsaw mills and ripping chains, is Bailey's Woodsman's Catalog (800) 322•4539. Or visit their website at: www.baileys-online.com.

The preceding suggestions are examples of the kind of creative approach you can take to solving the corner problem with post and beam framing. Based on your chosen geometry and the thickness of the wall, draw a full-scale section of the required post shape on a piece of cardboard and figure out a post design that makes good use of available materials. Don't trust my ideas. Test it out at full scale for yourself.

On Using Round Posts

Whatever design you create should offer a good support to the cordwood wall panel. In my view, a simple round post (shown in Image 7.6) does not do this. The wood will shrink away from the mortar, leaving a very weak unsupported column of mortar. Even bent nails

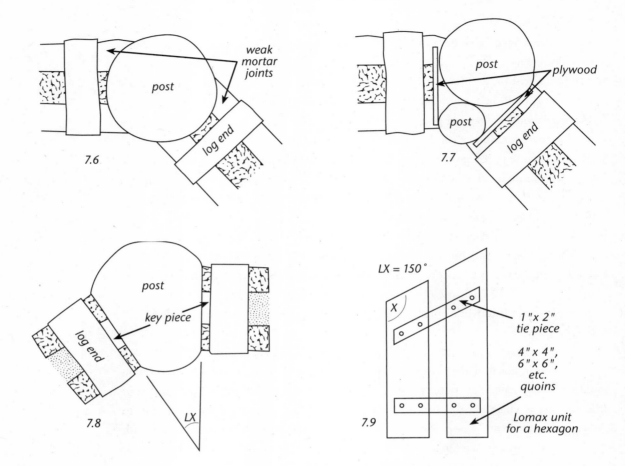

7.6, 7.7, 7.8 and 7.9
Credit: Rob Roy.

sticking out of the round post, which some people have used in an effort to supply a friction bond, is not an inherently good system. I'd rather do something different (as seen in Image 7.7), even though the curved part of the post sits quite proud of the cordwood masonry. That might look quite nice, in fact.

If you have really big trees, or if your cordwood walls are not all that thick, you can saw or hew two sides of the tree trunks to make posts (see Image 7.8. This is like a rustic version, really, of Image 7.1). Just figure out the angle you need (angle X in the diagram) for the polygon you choose. Angle X is determined by taking the total internal degrees of the polygon, dividing that figure by the number of sides to get supplementary angle Y (not shown on drawing), and then subtracting Y from 180. The following chart may help:

Regular Polygon	Number of Sides	Internal Degrees	Y (degrees/sides)	Angle X (180°–Y)
Square	4	360	90	90
Pentagon	5	540	108	72
Hexagon	6	720	120	60
Octagon	8	1080	135	45
Decagon	10	1440	144	36
Dodecagon	12	1800	150	30

Stackwall Corners and Polygons

Some builders have created multi-sided buildings with stackwall corners of regular crisscrossed lumber, but again, the advantage of building under cover is lost. If you are thinking of taking this route, consider designing some adaptation of the Lomax corner units described in Chapter 5. Image 7.9 is a plan view of a Lomax unit for a hexagon home. The tie pieces can be made from 1-by-2-inch stock, and the quoins can be any regular milled timber appropriate for the wall thickness, such as four-by-fours or six-by-sixes. Make half of the units about six inches longer, so that they can be staggered when placed in the corners, thus knitting the stackwall corners more strongly into the regular cordwood masonry.

Lomax units would probably work best with pentagons and hexagons, although Tom Kwiatkowski of Plattsburgh, New York nailed up stackwall corners of 2-by-6-inch and 2-by-4-inch material for his 12-sided cordwood home (see Image 28.1 in Chapter 28).

CHAPTER 8

Bottle Designs in a Cordwood Wall

Valerie Davidson

BOTTLES AND OTHER GLASSWARE can be used to create stunning effects in your cordwood walls. Patterns may be abstract or representative of a motif or idea that is important to you. The possibilities are endless. The following are some tips to help unleash your potential as a creative artist specializing in "conservation-minded stained glass."

Begin by collecting an array of colored and clear glass materials. An amazing array of glass materials can be suitable for use in cordwood walls. In the walls at Marshwood, our round two-story home in Parson, British Columbia, we have used liquor bottles, wine bottles, pop bottles, glasses, vases, glass electrical insulators, ashtrays, bowls, cosmetic containers, candy dishes, canning jars, and curved glass pieces from a lazy Susan.

Material for "bottle-logs" — Rob calls them "bottle-ends" — can be found inexpensively in many different places. Yard sales, thrift shops, discount bins at stores, neighbors, glass and bottle recyclers, dump sites, ditches, and alleys have all yielded glass treasures. Unless they are very thin glass, most articles that transmit light can be used. Some items were collected just because they were an interesting color or shape. Later, inspiration would strike, suggesting a pattern incorporating that piece. Keep in mind when collecting pieces of unusual shape that you will need a clear piece of a similar shape to pair with it to form a bottle-log.

Other things to consider:
1. It is best to place the colored piece on the inside of the wall, with a clear piece on the outside to maximize the amount of light transmitted through the bottle-log. Also, the vibrant color will now be on the inside, the normal vantage point for viewing. If the clear bottle is on the inside, and the colored one on the outside, then the color as seen from inside is very much diffused. However, when I've been unable to locate a bottle with the same color density as the others in a pattern, I've used two paler colored bottles to form a bottle-log. This helps to increase the color density of the log and minimize the variation of color density within the pattern.

8.1: Rainbow design.
Credit: Jim and Val Davidson

2. Bottle patterns with the same density of color in each of its pieces are easier to photograph.
3. When creating a pattern where all the bottles of one color are not exactly the same, e.g. the blue bottles in the "Celtic Cross" pattern at Marshwood (see the Color Section), try to distribute the different shades so they don't form a color blotch within the pattern.
4. The thicker the glass the better. Insulation is improved and there is less likelihood of it cracking or breaking during extreme temperature changes.

The property on which we are building Marshwood is an old homestead. No one had lived on it, prior to our purchase, for more than ten years. Friends of the previous owner and others had used the property to store unwanted items in the barn and other outbuildings. Included were several boxes of bottles in a collapsing garage — wine and liquor bottles primarily, including several Canadian Club (curved, amber) and tequila (curved, clear) mickey bottles. When I first saw them I almost threw them out, thinking they were the wrong size and shape.

Inspiration for our stained glass came more often on-the-spot than during advance planning. The first design was partially prompted at a cordwood workshop that Rob Roy was conducting at our property in 1996. Rob said, "Put the bottles at eye level." The theater director in me whispered, "Whose eye level?" Our daughter was expecting at the time, and a couple of weeks later our first granddaughter arrived. Suddenly these two events meshed with the almost-discarded CC and tequila bottles, and a "flower" was born (see the Color Section). Once the flower was designed, it became apparent that more bottles could be used to create a more complete scene. Hence green bottles for the grass, stem, and leaves; and blue bottles for the sky. Oh yes, and I must find a yellow glass for the sun ....

Also in the old garage, our son found a large log exactly the same length as the width of our walls. It inspired our "Madonna" wall (see the Color Section). Serendipity can play a large part in creating bottle designs.

There are several things to be taken into account when planning bottle designs: location within the wall, type of room, light and shadows cast by the sun, and artistic principles. Since we were working with an established pattern in the cordwood wall (explained in Chapter 9), fitting the glass into the walls was often fairly straightforward. A good example is the "rainbow" design (see Image 8.1).

The design principle of balance was addressed once the patterns were decided upon. First, we sketched the proposed pattern to scale. Then an imaginary frame was drawn around the portion of the wall that was going to contain the pattern. We considered several questions before proceeding: What, if any, furniture would be near the design? Would the placement of nearby doors and windows become visual design elements that might enhance or compete with the design? In what type of room would the design be located? Finally the bottle design was drawn in and moved around until it was balanced within itself and with the rest of the room.

For larger, more complicated designs, I highly recommend the creation of a full-scale drawing of the portion of the wall in which the bottles will be placed. Include the nearest architectural feature on each boundary, e.g., the nearest door or window to the left and right; the ceiling and floor; and any large piece of furniture covering any significant portion of the wall. This helps in visualizing the relationship of design size to room size.

Cordwood stained glass can be worked into any room in the house. At Marshwood, almost any wall that doesn't have a picture window has a bottle picture. (The view of the Rocky Mountains is better than even the best man-made works of art.)

When deciding where to put your bottle art in the exterior wall, it's helpful to consider the angle and location of the sun at various times of the year. Also, consider the number and type of pieces in the design. The "fish" design in our second floor bathroom is lit best in spring and autumn near the equinox (see the Color Section). At these times of the year, the setting sun is directly behind the design and its light comes in under the eaves of the house. The midwinter sun low in the south strikes our "flower" wall during most of the morning, sending bright splashes of color streaming across the floor and giving a lift to the short days. The "star" wall on the north quadrant of the house receives the best light late in the evening near the summer solstice (and early in the morning, I am told).

Many of the bottle pieces in Marshwood are brightly colored to capture the full impact of the sun behind them. However, we got a pleasant surprise when we discovered that the clear bottles in our "Madonna" wall — which is in the north-east quadrant of the house and seldom receives direct sunlight — turn the most wonderful copenhagen blue at dawn and dusk, a wonderful result of atmospheric light at these times of the day.

Now for the practical aspects of cordwood bottle patterns. You will need to find a source of aluminum printer's plates to form the bottle-logs. If you can find them, printer's plates that have not been inked are best. Sometimes the plate is exposed during shipping and becomes scrap. Ask at your local newspaper office or at a print shop. Although it is not a solution that I like (because of the volatile gases), Rob says that the printer's ink is easily removed with a little kerosene on a cloth rag. With the sketch of your stained glass pattern at hand, choose the bottles or glassware that will be placed on the inner side of the house wall — normally the side from which you will enjoy the design. Match each piece with another of approximately the same diameter. This will form the other half of your bottle-log.

All the bottles and glassware should be clean and dry. It's best to start with the cleanest pieces you can find. However, if bottles are a rare shape or color but are very dirty, try putting baking soda in them, then add pickling vinegar. Let the mixture sit for several minutes to several hours after it finishes bubbling, then scrub it with a good bottle brush. It is best to dry the bottles several hours in advance, especially during humid weather. Moisture in the bottles may lead to cracking or mould growth.

Work with a bottle-log length guide (cheat stick) the same length as your cordwood wall thickness. Lay the two pieces beside the guide to check their combined length. The best light transmission seems to happen when the bottle-ends protrude from the wall about a half-inch on both sides. If the two bottles are too short, but have good cylindrical shape, they will work fine if you follow the assembly steps listed below. If the bottles (placed neck to neck) are too long, the best thing to do is find a shorter companion bottle to be used as the outside piece. However, if (for some reason) you need to use the bottles at hand, several types of bottle cutters are available. It would be wise to practice on easily replaced bottles until you get the knack of using a cutter.

Assembly of Bottle-logs

Have at hand your bottles and other glassware, printer's plate, wide elastic bands, kitchen scissors, masking tape, silicone sealer, and a ruler or wall-thickness guide.

To create a simple two-bottle log, measure and cut the printer's plate. Depending on the width of your wall, leave at least two — and preferably three inches (five and preferably eight centimeters) — of the bottles exposed at each end. The wider the wall, the larger and heavier the bottles, and the greater the space between the bottle necks, the more contact you want between the bottles and the mortar.

Next, slip a wide elastic band around the bottom of each bottle. Then tear a piece of masking tape approximately three inches long and attach it, lengthwise, to one side edge of the printer's plate. Now, lay the bottles neck to neck and check their total length. Make sure it is no more than an inch longer than the thickness of your wall.

Place the first bottle on the printer's plate. Roll the plate around it until it is nice and snug. Hold the bottle and plate with one hand and use the other to press the masking tape down. Slide the elastic band you slipped on the end of the bottle up over the end of the plate. Turn the resulting tube around and slide the second bottle into the other end. Again, slip the elastic band over that end of the printer's plate. Adjust the masking tape, if necessary, to hold the center of the printer's plate firmly in place.

When wrapped around the bottles, the printer's plate should overlap itself about an inch. Besides making the bottle-logs more stable and easier to handle, the printer's plate holds the insulation away from the necks. Also, the plate reflects the light within the tube, allowing the maximum transmission of light.

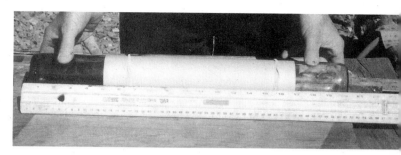

If your elastic bands are strong, you can often leave out the masking tape. Make what Rob calls a "spring-loaded, bottle-end cylinder" by installing an elastic band about an inch or a bit more from each end of a prerolled aluminum cylinder. Create a cylinder about a half-inch in diameter less than the bottles you are going to use. Then simply plug the necks into this "spring-loaded" cylinder (see Image 8.2). Very often, the resulting bottle-log is quite strong and stable without the use of tape. But keep them horizontal, just to be safe. The bottle-log is ready to be mortared into place, just like a log-end (see Image 8.3).

If both bottles are approximately the same size and cylindrically shaped, the elastic bands and the masking tape should hold them firmly together. However, it is still prudent to be careful when handling bottle-logs so that a bottle does not fall out and break while being mortared into place — it might be a rare or hard to replace piece.

To create a bottle-log using odd items (such as the blue, shell-shaped candy dish with a flat top edge that we used in the "fish" wall), first attach the item to the bottom of a similarly sized bottle, using clear silicone caulking. Keep the silicone to the outer edges as it may be visible through some items. Let the silicone dry as recommended by the manufacturer. Then make the bottle-log as usual.

Creating a Bottle-Log with Elaborately Shaped Glassware

First, decide how much of the item you want to protrude from the wall. Then select two suitable bottles to match the size of the item. Their length combined with that of the item should not exceed the width of the wall plus any planned protrusion.

Consider how the item will fit into or against the bottle that forms the inner portion of the bottle-log. (Remember that the light will travel through the bottle-log and light the

8.2: A colored bottle (left) and a clear bottle (right) are plugged into an aluminum cylinder "spring loaded" with strong elastic bands. This bottle-log is for a 20-inch-thick (50-centimeter-thick) wall. Plenty of glass is left exposed to bond with the mortar.

8.3: The first course of bottle-logs in a "poor man's stained glass panel" made by the Roys. The 20-inch wide (50-centimeter) wide panel has inner and outer 4-inch (10 centimeter) mortar joints and 11 inches (28 centimeters) of sawdust insulation.

8.4. Credit: Val Davidson.

special item primarily where it is in direct contact with the bottle.) The more irregularly shaped the item the trickier it is to attach.

Now figure out a way to balance the item on the bottom of the interior bottle. Silicone the item to the bottle and let it dry.

Cut the printer's plate to the correct width to go around the two bottles that form the main barrel of the bottle-log. Depending on the size of the item attached to the bottle-end, cut the printer's plate longer than usual for a standard bottle-log.

Cut into one end of the printer's plate to create tabs that will fit over the bond between the bottle and the item. Assemble the bottle-log as described previously, fitting the tabs firmly across the join between the special item and the bottle. Silicone the tabs to the item only where they will be covered with mortar, and trim the tabs to keep them hidden in the mortar. You may need to hold the tabs in place with wide, clear packing tape until the silicone dries. Then remove the tape. Be sure to leave enough glass surface exposed between the tabs to bond with the mortar.

To put our fish-shaped wine bottle in an "underwater" design, I had to figure out how to put it in the wall sideways. While contemplating the matter and sorting through my bottle collection, I came across some long bottles. By putting two of these bottles together using silicone, I had a base to use for the outside wall (see Image 8.4). It took some fancy balancing to get the fish bottle poised on the two necks and the tabs attached to it, but it was worth the effort.

Once your bottle-logs are all assembled, it is time to mortar them into the wall. For complicated designs, a template is highly recommended. It will not guarantee perfection, but it definitely improves the odds.

Take extra care when placing the bottle-logs in the wall. The best light transmission happens when the bottle-log is level, allowing the light to travel directly through both bottles. It is easy for one or the other of the bottles to get out of alignment during mortaring. Keep checking to ensure that the bottoms of the bottles are perpendicular to the floor, particularly if you will be viewing them from adult eye level.

Sometimes the primary view of the bottles may be from a specific location or angle (such as in a loft or stairwell) that is not in direct line of light transmission. There is a small degree of latitude in angling the inner bottles so that the bottoms of the bottles are better seen from this point of view. When you angle the bottles, check that all the bottle-ends in one pattern are flush with each other.

A chemical bond forms when the silica in the bottle glass fuses with the cement in the mortar. Once the mortar is dry and the bond complete, the bottle-log will remain firmly in place, and without the wood shrinkage cracks which sometimes occur with log-ends.

Once completely dry, the bits and pieces of mortar on the protruding glass are extremely difficult to remove. Be careful not to apply too much moisture while cleaning the bottle-ends as the mortar is drying, as this will adversely affect how the mortar sets. It seems that the best time to clean the bottles of mortar is a few hours after putting them in place. A dry cloth works as well as anything, although it can be moistened slightly, if necessary. Keep an eye on your bottle-logs as you complete the cordwood masonry above them. They are almost certain to get dirty again — consider masking them off until the masonry is complete.

Should the mortar dropped on the bottle-ends become too dry to remove with water, try scrubbing with pickling vinegar. If this is not strong enough, you could use muriatic (hydrochloric) acid, which can be purchased at a hardware store. Or try rubbing the mortar-splattered bottles with fiberglass insulation. Be sure to wear gloves when cleaning bottle-ends by any of these methods.

I hope you find this information useful, and that you enjoy creating and viewing your stained-glass art work for years to come.

CHAPTER 9

Patterned Cordwood Masonry

Rob Roy

VERY SOON AFTER ARRIVING at Val and Jim Davidson's beautiful 80-acre (32-hectare) lot along the Columbia River in Parson, British Columbia, Jim took me to look at the cordwood we'd be using during a three-day workshop scheduled to begin a couple of days later. My jaw must have visibly dropped as Jim showed me two different tidy stacks of extremely regular cylindrical log-ends, all cut precisely to 16-inch (40-centimeter) lengths. "This pile," he told me, indicating the rank of smaller log-ends, "is cut from the spruce peeler cores left over at a local plywood plant." Each log-end was precisely 3½ inches (89 millimeters) in diameter. Peeler cores, I learned, were the 8-foot (2.4-meter) cylinders left over when larger logs are veneered at the plywood plant. I presumed that 3½ inches is the minimum useful diameter, beyond which the veneering machine can't peel any further.

"Uh-huh," I said, "and what about this other pile?"

"Oh, these I got from a local fence post supplier," said Jim. "They're mostly pine."

"Uh-huh," I repeated, lost for words. Virtually every one of the larger cylinders in this second stack was within a half-inch of 7½ inches (19 centimeters) in diameter. I had no idea how I was going to teach a cordwood masonry class with exactly two different log-ends to choose from. At every other workshop I'd conducted over the previous 20 years, there had always been a good variety of sizes and shapes of log-ends available. Jaki and I would teach what we call the "random rubble" pattern of construction. No way I could do random rubble with this! My mind was racing.

"Well, how do you like the wood?" asked Jim.

"It seems ... well, it's very dry, no bark on it ... and it's very ... *regular*," I stammered.

"But?"

"But I can't imagine how we are going to use both kinds of wood in the wall. I know that Cliff Shockey in Saskatchewan has built a number of buildings with predominantly one size of wood, recycled cedar utility poles, actually. He lays them very quickly in regular courses.

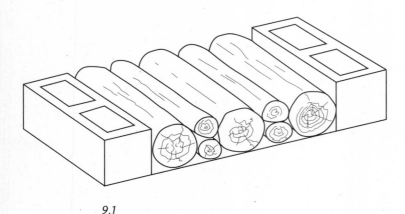

9.1

A kind of hexagonal configuration evolves, like the walls of a honeybee's nest. I just don't know what we're gonna do with these little guys."

There were some concrete blocks lying around on the slab that Jim had already poured. He had not begun any of the cordwood work, wanting to learn the technique himself at the workshop that he and Val were hosting. I placed a couple of blocks on the foundation, transversely on the footing like log-ends, and used them to support a little sample construction (see Image 9.1).

Jim watched me at work. I don't think that it ever occurred to him that there could be any kind of problem associated with having such perfect log-ends. The little structure took about five minutes to figure out. I showed it to Jim and later to Val.

"This is the only way that I can think of to use both sizes of wood," I said. "And it looks like this pattern will deplete both sizes of wood at about the same rate. We'd be using exactly twice as many small log-ends as big ones. And the mortar joint will actually improve the appearance of the pattern, I think. The second course can be placed with the large logs over the small, and vice versa. The work should go fast. We'll never need to search for the right log-end." Jim and Val approved the pattern and the house was built that way.

I still had three concerns about the pattern. My first was that it might be hard to keep up the pattern with good, even regularity. I'd long known that two ways of getting a poor looking wall were to try to build a random wall, only to have it evolve into an unwanted pattern; and to try to do a pattern but be unable to keep it up because of depleting a certain size of log-end. Well, we would not run out of either size of log-end at the Davidson's place. They had plenty.

My second concern was how the pattern would work up against door and window frames and also how it would fit in under windows and at the first-floor joist level.

In the event, my second concern did not become a problem, either. I was very careful to impress upon the workshop students that it was important to keep the courses level. Constant mortar joints were even more important here than with the random rubble style. We even built a little pattern drawn out on plywood — something like the "idiot stick" I use to get windows and joist plates all at the correct height — and this proved useful, particularly on the first day, in keeping the courses rising at a consistent rate. Also, we always started work at a door frame. If there was a second door frame, or a tall window frame nearby, we'd measure and plan the spacing of the logs so that they would come out right. Usually, this would involve no more than a quarter-inch adjustment to the width of each vertical mortar joint, unnoticeable to the naked eye. On every second course, next to a door or window frame, it was necessary to split either a large or two small log-ends in half, in order to fill the space and avoid huge mortar joints.

My third concern was that the wall would be, well, boring. It turned out to be anything but. The pattern has all sorts of unexpected geometries tied up in it. You can find rectangles, diamonds, and hexagons in the pattern. And Val's creative bottle-end art makes the Davidson house one of the most visually interesting cordwood structures ever built.

9.2: A patterned cordwood wall.

Cliff Shockey was present at the workshop as a guest instructor, and he, too, was impressed with the wall's appearance, as are most people who visit the home or look at pictures of it in our photo albums.

In fact, Val's artistic conservation-minded stained glass designs are all the more impressive when one considers that she had to knit her designs into a rigidly patterned wall.

One evening during the workshop, Val asked me if I thought it was possible to put a Celtic Cross of bottle-ends into this patterned wall. I remember sitting around the little travel trailer they were living in and doodling up a design on a piece of scrap paper. The large log-ends — the 7½-inch (19-centimeter) guys — were about the same diameter as gallon wine jugs. And the little peeler cores were the same size as certain wine and beer bottles. I helped the Davidsons design a Celtic Cross that would fit in with their cordwood pattern, and Val and Jim executed it perfectly after the workshop was finished (see the Color Section).

Jaki and I saw the cross two years later when we returned to the Davidson's idyllic corner of the world to conduct another workshop. I cannot begin to describe the vibrancy of the blue and green light that enters the home through this particular design. And Val, as the reader knows from the previous chapter, took bottle-end art to even greater levels of creativity.

So if you are cursed or blessed — they are the same in this case, just a state of mind separates them — with only one, two, or even three distinct sizes or shapes of log-ends, take a few minutes or hours to work out a pattern that makes visually pleasing and efficient use of the material that you have available. Remember, lemons make good lemonade.

The Davidson's log-ends were not only regular in diameter but were also very precisely cut for length and for plumb by a professional woodcutter with a crosscut saw. Quality log-ends of this type are very nice to work with and yield a more finished appearance. Many cordwood builders, such as Larry Schuth (see Chapter 17) take the time to build a cordwood cutoff table for their chainsaw.

Making a Cordwood Cutoff Table for a Chainsaw

(Editor's Note: The remainder of this chapter is an edited version of an article that originally appeared in The Mother Earth News *(May-June, 1982). Over the years, many people have used this design or an adapted version of it to make an excellent tool for cutting cordwood precisely and safely. (See pictures included.) The original prototype was conceptualized and built by Bill Weiner in New Brunswick back in the '70s, in collaboration with Cordwood Jack Henstridge. Jack was still using the original as of 2002. Here, we describe a successful model made by technicians at the former Mother's Eco-Village in North Carolina.)*

For small cordwood projects, it is perfectly feasible to cut all your log-ends by hand with your chainsaw. Just measure, mark, and cut. Use a saw bench for safety and have a trusted and careful partner hold the log steady for you, while keeping well away from the saw's chain. But, for a large cordwood project, like Jim and Val's house, it is worthwhile to use a farmer's buzz saw made for the purpose, or to make a cutoff table as described here. The schematic drawing (see Image 9.3) has all the information any good backyard tinkerer would need for building the cutoff table, but here are some additional tips:

The *Mother Earth News* model has its tabletop made from some 2-by-6-inch tongue and groove pine decking that was lying around, but any two-inch-thick lumber would do. Just be careful to locate the carriage bolts no more than six inches from the working end of the top, to assure that the saw chain cannot possibly hit one of them. *Mother's* work crew made an adjustable length stop, so that every log-end would be precisely the same length. They made theirs from a piece of ¾-inch, schedule 40 pipe, which slides nicely inside a section of 1-¼-inch box tubing. But, as it happens, the ¼-inch pipe will also slide snugly inside of a 1-inch, schedule 40 pipe. Whatever combination you decide to use, be sure to include the bent ⅜-by-3½-inch carriage bolt that serves as a setscrew.

For a pivot mechanism for the chainsaw itself, there is a temptation to use a similar make-do bushing arrangement as was used for the length stop. *Mother's* technicians tried that approach themselves, but opted for the "pillow block" setup (see Image 9.4). The ball

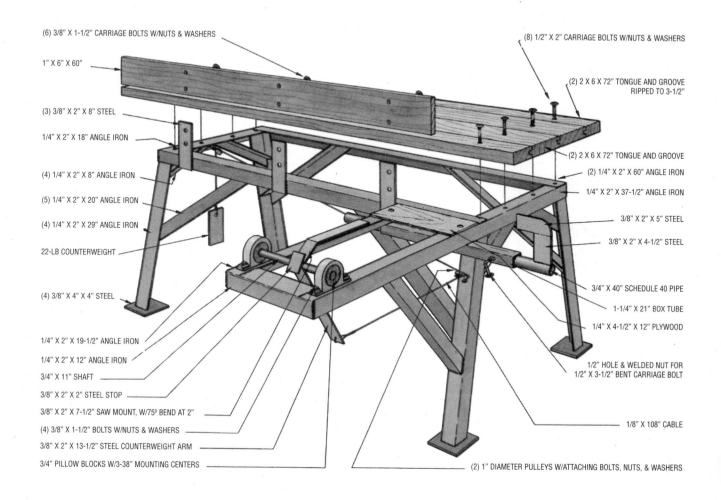

bearings of the pillow block help the saw to move smoothly through its arc and to keep the blade rigidly square to the log.

Of course, when you first bolt your saw's handle to the pivot arm assembly, you may find that the bar isn't vertically square to the table. In order to get their Stihl 048 chainsaw to run perfectly true, the crew sandwiched a 3/16-by-3-by-12-inch piece of steel between the pivot arm and the handle (both to provide extra rigidity and to compensate for irregularities on the saw's underside) and then shimmed the pillow blocks as needed.

Fastening your chainsaw to the pillow block assembly is the part of this project that will probably require the most creative backyard engineering, as every chainsaw is different, and none of them was designed for the purpose. Larry Schuth actually used a very heavy hinge

9.3: Cordwood cutoff table. Credit: The Mother Earth News. *May/June 1982. p. 110.*

in place of the pillow block assembly and used bolts, metal plates, and rubber shims to creatively and firmly mount his chainsaw to this hinged pivot assembly.

Mother's design calls for a cable and pulley counterweighting system, and they strongly advise that you go to the trouble of including it. This arrangement ensures that the chainsaw will remain in whatever attitude you set it. For example, you can raise the saw and leave it up for log-loading. And the counterbalancing prevents the saw's weight from carrying it down into the wood. Instead, the operator controls the speed of the cut by applying pressure to the handle.

9.4: Pillow blocks make an excellent pivoting mechanism.

9.5: Charles Shedd of Bakersville, North Carolina used his ingenuity to build this very effective — and portable — cordwood cutoff mechanism.

Charles Shedd made an adaptation of the cutoff table out of galvanized pipe. The pivot mechanism was simply a short length of male-threaded pipe that could turn within the female threads of adjacent elbows. Charles achieved a counterbalance without the pulley system by mounting his Craftsman chainsaw to a long plank of wood, gaining great mechanical advantage which translates into ease of operation.

Several features in *The Mother Earth News* design are specifically intended to increase operator safety. Because the saw pivots from the handle, wood chips are thrown at the ground, not toward the user's face. And, in the unlikely event of chain failure, the cutting links would be directed away from the operator's body. In addition, the direction of the chain travel pulls the wood into the backboard to prevent skipping and to lessen vibration. Still, it's a good idea to have an assistant brace the far end of the log to prevent it from twisting.

Mother's technicians found that the minimum size saw for effective use with their cutoff table was 2.5 cubic inches (41 cubic centimeters), with a bar of at least 16 inches (40 centimeters). In fact, they really preferred a saw of between 3.5 and 5 cubic inches (57 and 82 cubic centimeters), with a 20-inch-long (50-centimeter-long) bar.

With chainsaw operation, kickback is the primary hazard facing the user and is usually the result of the moving cutting chain at the tip of the bar coming into contact with an object. With the cutoff table, the tendency of the chainsaw to pull the log against the backboard helps ward off the possibility of kickback. But, as added insurance, the designers built in a stop to prevent the saw from pivoting beyond about a 45-degree angle. Whatever adaptations you make to the table design, make sure that the saw safely "bottoms out" before the chain can touch anything except the log it was designed to cut.

Only use a saw with a chain brake feature. The brake will stop the chain in the event of kickback. Also, the brake can be used to lock the chain while you position a log on the tabletop. Bruce Kilgore of Morrisonville, New York cut all the cordwood for his home while working alone. Without a chain brake, the only safe way to load the cordwood cutoff table is to shut the motor off while doing so.

Incidentally, Bruce had difficulty finding the pillow block assembly but finally found that they were available from Lee Valley Woodworking Catalog at: www.leevalley.com or (800) 871•8158 and through Woodcraft at (800) 225•1153.

For safety, quality control, speed, and ease of operation, it is well worth taking the time to marry your chainsaw to a cutoff table. It may require some improvisation, but the extra effort will pay great dividends.

9.6: The Mother's Cordwood cutoff table. Credit: The Mother Earth News. May/June 1982. p. 110.

9.7: The table in action.

CHAPTER 10

Electrical Wiring in Cordwood Masonry Buildings

Paul Mikalauskas and Mike Abel

(Editor's Note: Most of this chapter was written by the late Paul Mikalauskas, builder of "Earthwood Junior" in Ashland, New Hampshire. He presented his paper on the subject at the 1999 Contintental Cordwood Conference in Cambridge, New York. A friend to all, Paul succumbed to cancer in 2001. The sidebar is by Mike Abel, who built a beautiful cordwood home in Wetherby, Missouri, where he is a licensed electrician.)

ELECTRICAL WIRING IN CORDWOOD MASONRY BUILDINGS presents challenges different from wiring in conventional stick frame construction. One functional difference is that many cordwood buildings do not have a basement in which to hide wiring. With a little planning, however, wiring does not have to be very much more difficult than with other building styles. With creativity, one may find many nooks and crannies where wiring can be hidden.

A current copy of the *National Electrical Code* is an excellent investment. The code book will make it easier to assure that the work is both codeworthy and safe, and will give the reader answers about wire and conduit size, how many conductors will fit in a box, and so forth. The code promotes safe wiring practice, good for owner-builders, as well as for any future occupants — thinking about possible resale is not a bad thing.

The service entrance is where the electrical power first enters the building. In many conventional homes, the power company simply stretches a line from the primary pole to a conduit on the house that serves as a mast for the incoming power. Heavy gauge wires run down to the service entrance panel, often in the basement. This panel contains a main breaker and individual circuit breakers for the various lighting and small appliance circuits in the home.

10.1: Power is brought underground to the service entrance at this suburban cordwood home in Eau Claire, Wisconsin.

10.2: Electrical boxes can be supplied by flexible conduit running within the insulation cavity. Credit: CoCoCo.

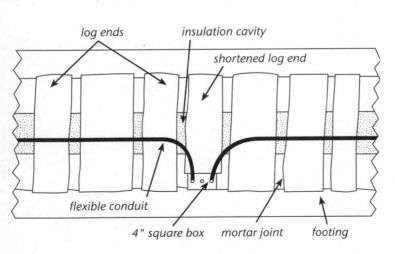

Utilities that enter the building from underground, however, may be more in keeping with the natural appearance of cordwood buildings. If the house is to be built on a concrete slab, the builder will need to run correctly sized conduits in the earth and below the slab before the slab is poured. This is also the time to identify and accommodate for any freestanding features in the building not accessible from above or from an interior wall — features such as kitchen islands, a duplex receptacle (wall plug) near a masonry mass, and the like. Branch circuits may be run to the proper locations using schedule 40 PVC conduit.

Thought should also be given to any future needs for outdoor power away from the building, and conduits should be run underground to the point or points where they may be used when needed. Conduits may also be run to outlet and switch locations in the insulation cavity of the cordwood wall. Install elbows to place electrical boxes flush with the finished interior wall.

Wiring for wall outlet circuits may be laid in the insulation cavity during wall construction. Flexible wall conduits are recommended for this (see Image 10.2). At least one cordwood builder, however (Ed McAllen of Galesville, Wisconsin) used direct burial Romex™ conductors in the center of his 16-inch-thick (40-centimeter-thick) cordwood walls and met with code approval because the Romex™ was always more than 4 inches (10 centimeters) from either surface of the wall. He brought the conductors into the back of his electrical boxes, which were set flush into large log-ends. During the winter prior to building, Ed prepared 20 or so 10-inch diameter logs for this purpose, by cutting and chiseling correctly sized rectangular openings into the logs to receive the boxes. He routed a pathway from the box opening to the center of the log to carry the Romex™ from the insulation cavity into the back of the electrical box.

Cliff Shockey employs a similar detail with his double wall technique. After building his outer cordwood wall and installing the hardboard, insulation, and vapor barrier in the middle third of the wall, Cliff runs his rough wiring for interior duplex receptacles, switches, and lights. During construction of the inner wall, special notched log-ends, similar to McAllen's, are placed where they are needed according to the electrical plan. The rough wiring is brought into the box, leaving eight to ten inches extra for making final connections later.

There may be sections where it is not possible or desirable to hide the wire in the cordwood wall or under the floor. In those cases, wiring may be enclosed using Wiremold™ or electrical metal tubing (EMT) conduit on the interior cordwood wall surface or along posts, beams, and window or door frames. Exposed conduit or Wiremold™ is code approved and has several advantages for the cordwood masonry builder: Using this method, cordwood masonry production is not further slowed by taking time to weave conduit or Romex™ through the insulated cavity. Electrical can be installed after the cordwood walls are built and the roof installed. Also, the electrical circuits are readily accessible to facilitate changes, repairs, or additions.

10.3: Surface-mounted Wiremold™ allows the electric to be installed after the walls are built.

There are some disadvantages to surface-mounted wiring. The Wiremold™ or EMT adds extra cost to the electrical component. New skills must be learned to make a nice job of surface-mounted wiring. And some people may not like to see surface-mounted conduit, although it is becoming more common all the time, particularly in commercial buildings. Wiremold™ (and other available systems) comes in a variety of colors and EMT conduit can be painted to match or contrast. By careful planning and intersection with interior partitions (where conventional wiring practices may be used), it is possible to minimize the amount of surface-mounted wiring quite a bit, although code does require a duplex receptacle every 12 feet (3.6 meters) around the perimeter of all rooms.

Feeds may be run from the distribution panel to points around the building by using the space left between the inner and outer wooden plates, often made from 2-by-6-inch planking, at the top of the cordwood wall, if your construction method happens to incorporate that detail. Wiring to lighting fixtures can be run along the top side of girders, if exposed post and beam construction is used in the home. If you build up your own box posts for a post and beam frame using, for example, two-by-sixes and two-by-tens (see Image 10.4), then wiring can also be run inside the box post cavity. (See also Chapter 6 for how Bunny and Bear Fraser cleverly incorporated switches and receptacles every 8 feet (2.4 meters) around a 16-sided round post and beam frame.)

10.4: Box post made from 2-by-6-inch and 2-by-10-inch lumber. Use screws or coated nails and wood glue.

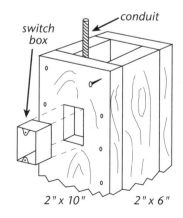

Other builders have run a baseboard around the base of cordwood walls, incorporating conduit or Romex™ conductor behind the baseboard and surface-mounted boxes on the baseboard surface. If the first course of similarly dimensioned logs is cut an inch or two shorter than normal — 14 inches (36 centimeters) instead of 16 inches (40 centimeters), for example — the baseboard need not protrude into the room.

Electrical in Cordwood: Some Additional Comments
Mike Abel, Licensed Electrician

In my round cordwood home, I used a combination of several of the methods Paul mentions but relied most heavily on flexible metallic conduit (also known as "flex"), with individual, appropriately sized stranded conductors of type THHN insulation. From my rigid conduit stub-up in the slab at the exterior wall cavity, I changed to flex with the appropriate fitting and moved on to my wall outlets and switches. The flex snakes satisfactorily through the insulation cavity, and then, using a flex connector, terminates at the end of a shortened log-end (see Image 10.2). While building the wall, I determined the length of the flex needed, cut it with a hacksaw, and then used a fish-tape to pull a pulling string, to be used later to pull in the wire. It is important to put the string in before embedding the flex in the wall, as it is essentially impossible to send a fish-tape through the flex later. Use a multistrand poly pulling string, similar to baling twine — it is quite strong — and a wire-pulling lubricant, such as Ideal's Yellow 77, to lube the wires. The flex method is far superior to direct burial or NM (Romex™) in the cordwood walls, as greater flexibility is gained at the time of the installation, as well as later when electrical changes may be desired.

For switch and outlet boxes, I used 4-inch-square (10-centimeter-square) metal boxes, which provide more room for wire pulling and for the making up of connections. These can be purchased in standard 1$\frac{1}{2}$-inch (4-centimeter) depth or deeper and can be extended in depth with extension rings. Also, this type of box allows a normal duplex receptacle location to become a double-duplex location. All of my outlets are double-duplex — an additional receptacle only costs about fifty cents — advisable because of the impossibility of getting into the walls later. For the same reason, I put those double-duplexes every seven feet (two meters) around the perimeter.

Flex needs to be grounded, as the NEC code book will tell you. Grounding is important. Prior to pouring the slab or foundation, an 8-foot-by-$\frac{1}{2}$-inch grounding rod should be sunk into the ground near the service panel location. In addition, most utilities will be using a grounded neutral system, and the neutral should be grounded at the transformer. Any readers who find this all rather technical should consult their utility company, the NEC code book, or a licensed electrician.

In the slab rough-in for my round cordwood house, I included two stub-outs to all four compass points for future uses. Already, I have used one to provide power to my woodshed 30 feet (9 meters) away, as well as to provide an interior three-way switch for the woodshed light.

If the builder desires to have backup power in the form of a generator, then this feature may be part of the original electrical plan, or it may be added at any time in the future. The homeowner can choose backup for only those circuits that are deemed necessary.

With powerful computers in many homes, thought should be given to determine any future locations for computers and related equipment, such as phone jacks. The wiring requirements for technological devices are changing and uncertain. For this reason, both RG-6 coaxial cable and Category 5 phone/data wire should be run from the utility room to any location which might receive a computer, phone, fax machine, or television. Consideration should be given to any speaker locations, and in-wall wire should be run from the stereo to these locations. These can be mounted by one of the methods suggested earlier and terminated with readily available wall plates. Burglar and smoke alarms should be considered at the design stage, with wiring run at the appropriate time during construction.

Just as much consideration should be given to home power systems wiring, such as solar, wind, and small hydro systems. In fact, the National Electrical Code now addresses many of the issues involved with independent power. You may or may not have to pass an electrical inspection in the case of homemade power, but it is a good idea to have it inspected anyway. The inspector may spot something that might save your building or your life.

No matter which wiring method you choose, make a good wiring circuit diagram to work from. This is your roadmap, and if you sub the work out to a licensed electrical contractor, he or she will insist upon it and check it for codeworthiness.

The author wishes to acknowledge contributions to this paper from Cliff Shockey and Rob Roy; cordwood owner-builder Ed McAllen of Galesville, Wisconsin; and especially to licensed electrician and cordwood builder Mike Abel of Wetherby, Missouri.

CHAPTER 11

Using Cement Retarder with Cordwood Masonry

Rob Roy

The Problem of Mortar Shrinkage

WHEN CEMENTITIOUS MATERIAL (mortar, concrete, plaster, etc.) dries too quickly, shrinkage cracks can develop. By slowing the set of the material, this shrinkage can be reduced. When pouring a concrete slab, for example, plastic is placed on the sand or gravel pad prior to the pour, which greatly reduces the transfer of moisture from the concrete to the ground. To further retard the set of the concrete — and to increase its strength — the slab can be flooded with water during the first day or two of the cure. Where full strength and absolutely minimal shrinkage is required, such as on a bridge deck, a commercially available retarder is added at the concrete batch plant.

Absorption Characteristics of Wood Masonry Units

With brick, block, or stone masonry, mortar shrinkage is not normally a problem. Masons will soak bricks, blocks, and porous stones so that these masonry units do not rob the moisture quickly from the mortar. Again, if the mortar sets too rapidly, shrinkage will occur and mortar strength is compromised. With cordwood masonry, however, soaking the masonry units — the log-ends — is a bad idea, with two likely negative results. First, the wood will swell prior to laying it up in the wall and, therefore, cause greater wood shrinkage gaps when it returns to its presoaked moisture content. Second, the wood might continue to swell while the wall cures, thus breaking up the mortar joint. Neither of these situations is a happy result, and the second can actually cause other structural problems with all three styles of cordwood masonry. With a round house, wood expansion can cause the wall to tilt outward. With cordwood used as infilling in a post and beam frame, there can be an uplifting on the plate beam at the top of the wall. And stackwall corners can be forced out of plumb.

Wood absorbs moisture from mortar quite rapidly, which can cause both wood expansion and mortar shrinkage. The variables at play in moisture absorption in wood are discussed more thoroughly in Chapter 22, as well as one method — the use of waterseal-type products — which we have used successfully to decrease the transfer of moisture. In this chapter, however, the emphasis is on avoiding mortar shrinkage.

Sawdust Mortar

At Log End Cottage, the dry cedar log-ends caused a rapid drying of our mortar, which was composed of sand, Portland cement, and builder's lime. We tried all sorts of methods to slow the set, including hanging damp towels over the work. Nothing helped until the last panel of the house, where we introduced soaked sawdust into the mix, in an attempt to slow the set. That worked. There were no shrinkage cracks in the sawdust test panel. The softwood sawdust particles, which had been passed through a half-inch screen and soaked overnight (or longer) in water, acted like thousands of little water storage units — little sponges, as it were. My theory is that these little reservoirs give moisture back into the mortar as it dries, thus slowing the set. Another theory is that the high lignum content of sawdust acts as a retardant. Lignum is one of the main ingredients in Sika Plastiment™ cement retarder. Our sawdust mortar mixes are described in Chapter 3. Please see also the compression strength test reports of two different sawdust mortars, which appear in Chapter 30.

Cement Types

You can always be sure of strength quality with Type I Portland cement, whereas masonry cement varies according to type. Examples: By proportion, Type N masonry cement is composed of 50 percent Portland clinker and 50 percent ground limestone. Type S masonry cement is 60 percent Portland clinker and 40 percent limestone. Type M is 75 percent Portland clinker and 25 percent limestone. Portland "clinker" is different from Type I Portland cement, and ground limestone is different from Type S hydrated lime, yet good mixes can be made with either Portland cement or masonry cement. But watch out for "mortar mix." One or two correspondents have gone to their building supply yards and asked for masonry cement but were given "mortar mix" instead. Mortar mix is composed of 1 part masonry cement and 2½ to 3 parts sand. Obviously, mortar "mix" cannot be substituted for masonry cement in cordwood mortar formulas. The mud would have a very weak cement component and the mortar would crumble. Straight mortar mix by itself would be a fairly good mortar for cordwood masonry, although something would have to be done about mortar shrinkage, such as adding cement retarder.

Variation in Sawdust Qualities

In our standard cordwood masonry mix, the soaked sawdust retards the set. Without it, the mix feels hard to the fingernail the very next day. The sawdust is also the ingredient in the mix that can vary the most in characteristics. The ideal sawdust is one that is light and fluffy, as opposed to hard and grainy. I have had good success with red and white pine sawdust, as well as spruce, white cedar, and poplar — the same woods, incidentally, that are less prone to swelling as log-ends. I have not had good results with dense hardwood sawdusts such as oak. These individual particles are more like eighth-inch cubes of wood than like little water-storing sponges. Sawdust from dense hardwoods simply makes a grainy, crumbly, and hard-to-work mortar, and it would be better to leave it out of the mix altogether.

The sawdust we use is the kind that comes from a sawmill having a large circular saw blade. I've also had good luck with the sawdust made by a chainsaw. I have never tested the finer bandsaw sawdust to see if it has the same mortar retarding characteristics. My gut feeling is that the finer sawdust would not store moisture as well as the coarser stuff, so there would be less moisture available to be given off into the mix as it tries to set. This view was somewhat substantiated recently by a builder in British Columbia, who did not have great success with very fine sawdust, and by tests we performed at Earthwood workshops in the late '90s. Also, I am a little doubtful about the efficacy of the fine sanding sawdust that comes from a cabinetmaker's shop, although it would be worth testing.

To test any mortar mix for shrinking, build a small section of cordwood masonry with the mix you want to test — say a panel 3 feet by 4 feet (1 meter by 1.2 meters) or thereabouts. Check it every day with the fingernail test: a slow setting (non-shrink) mortar is easily scratched with the fingernail the day after construction. It can still be scratched, although not as easily, on the second or third day. After three or four days, the mortar should be hard. If mortar shrinkage cracks are going to appear, they usually show up within a week or two. If, after two weeks, you have no cracks in the mortar between log-ends, you have a good mix. But keep in mind that cracks in the mortar can also be the result of wood expansion. To be sure on this point, build the test panel under cover and avoid the use of woods prone to expansion, as discussed earlier.

The paper-enhanced mortar (PEM) advocates in Chapters 14 and 15 have had good results (very few mortar shrinkage cracks) with their mixes, although the PEM takes much longer to fully cure than more traditional cordwood mortars. Harold Johnson of Ellenburgh, New York used wood pulp mixed with water as his admixture and also had good strong non-shrink mortar. The mortar was easy to work and did not set too quickly. (See Complete Book of Cordwood Masonry Housebuilding, pages 94–95.) Incidentally, planer shavings are not an effective admixture for cordwood masonry mortar.

Screen and Soak the Sawdust

It is imperative that the sawdust be passed through a half-inch grid screen and be soaked at least overnight. The screen removes leaves, grass, bark, bits of wood, and the like — stuff that you definitely would not introduce deliberately into your mortar. Soaking the sawdust overnight allows the sawdust to fully absorb the water.

The worst thing you could do would be to add dry sawdust to the mix. The dry sawdust would absorb moisture from the other constituent ingredients, thereby accelerating the set! I've had a few late-night phone calls from people who reported severe mortar shrinkage "even though I'm adding sawdust." Their sawdust, it turned out, had not been presoaked.

A Problem of Quality Control

As discussed, sawdust can vary tremendously. People call and ask me if catalpa sawdust will be good or cabbage palm or some exotic wood from Central America. I have no idea! Try that small test panel first before building permanent walls. You'll know right away if the mix is grainy and hard to work or smooth and easy to work. The fingernail test will tell you whether or not the sawdust is helping to retard the mix — particularly if you test it alongside a non-sawdust mortar laid up about the same time.

In many parts of the country, it is simply not possible to obtain light and fluffy sawdust, because only dense hardwood is locally available. Because of the vagaries of sawdust and the difficulty of assuring quality control around the world, I have long hoped for an alternative to the sawdust admixture, an alternative that would give the desired non-shrink characteristic and that could be standardized more easily.

Testing Cement Retarder at Earthwood Building School

I was inspired to test cement retarder by Hans Hebel of Chile (see Chapter 21) and Olle Lind of Sweden (see Chapter 22). Independently and on different continents, Hans and Olle experimented with different retarders. Hans built his entire house addition with Sika cement retarder and was very pleased with the results. Olle used Cemtex retarder with success. I decided to do some controlled tests at Earthwood during our 1999 workshops.

The May workshop project was infilling the post and beam frame of our garage with cordwood masonry panels roughly 4 feet (1.2 meters) high by 6 feet (1.8 meters) wide. The cordwood wall is 8 inches (20 centimeters) thick. The inner mortar joint was separated from the outer mortar joint by sawdust insulation, so the width of the mortar joints was roughly 2½ inches (63 millimeters), as was the sawdust insulation layer. The thickness of the mortar varied from about ¾ of an inch to 1½ inches (19 millimeters to 38 millimeters), with some joints thinner or thicker than this when log-end selection was not the best.

Cement retarder is not easy to find in local supply yards, particularly in rural areas. My usual supplier of cement products advised me that there was no demand for retarder by local masons. He was willing to order some for me, but I was in a hurry to conduct tests at the upcoming workshop, so I decided to try the local concrete batch plant. A friend at the plant arranged for me to get a gallon of cement retarder from their huge vat of the stuff. I soon came away with a gallon of Daratard 17, made by Grace Construction Products. In their product information sheet, Grace describes the product as an "initial set retarder meeting ASTM C 494, Type B and Type D". Also, "Daratard 17 admixture is a ready to use aqueous solution of hydroxolated organic compounds." Wow! Impressive and very official sounding. I also learned that the recommended addition rates range from 2 to 8 fluid ounces per 100 pounds of cement (130 to 520 milliliters per 44.4 kilograms of cement).

On May 30, 1999 we laid up two parallel panels (A and B) of cordwood masonry. In each panel, the log-ends were mixed; about half were spruce (rounds and splits), and the rest consisted of cedar and basswood rounds. All of the wood was seasoned at least two years. With test panel A, we used our usual mix of 9 parts sand, 3 parts soaked softwood sawdust, 2 parts Portland, 3 parts lime. Test panel B consisted of 10 parts sand, 2 parts Portland, 3 parts lime, and 4 ounces of Daratard 17 added to the mixing water. (I reasoned that an extra shovel of sand would offset the missing bulk of the sawdust.) Weather conditions were hot (90 degrees Fahrenheit [32 degrees Celsius]) and humid, but we worked both panels under the shade of the overhang. Both mixes had long working times of at least 1½ hours. I felt that the Daratard mix was easier to point, a view shared by some of the students, but Jaki (a better pointer than I) preferred the "body" of the sawdust mix for pointing.

By 5 p.m. of the following day, both panels were fairly easy to scratch with one's fingernail, although the sawdust mix seemed to be just slightly harder. The next morning, (June 1), the Daratard mix was still a little softer than the sawdust mix. No shrinkage cracks had appeared in either panel. As preliminary results were good, we decided to try a section of wall with the same 10-2-3 mix, but with 3 ounces of retarder instead of 4 ounces. As a fiscal conservative, I figured that we ought to figure out how *little* can be used. This test panel was laid up on June 2 at 2 p.m., a sunny day of 76 degrees Fahrenheit (24 degrees Celsius), although we still worked in the shade.

On June 3, we noticed six hairline cracks in the May 30 mix that had been made with 4 ounces of retarder. About the same amount and type of cracks also appeared in the sawdust mortar laid up the same day. These cracks opened just a wee bit more over the next day or two and then stabilized. (All of the mortar, by this time, was fully hard, like rock.) But the big discovery was that the batches laid up on June 2 (with 3 ounces of Daratard 17) showed no mortar cracking at all. Two months later, there were still no cracks.

At the July workshops, we continued our parallel tests. After a day or so using 3 ounces of retarder, we cut back to 2½ ounces. This seemed good, at first, but after a week or so, we noticed cracking in the sections of wall where we used 2½ ounces, but virtually no cracking where we used 3 ounces. My conclusion is that 3 ounces of Daratard 17 added to a mix of 10 parts sand, 2 parts Portland and 3 parts Type S lime results in a good, hard, strong, non-shrink mortar, equal or superior to the sawdust mix made with the best possible sawdust. I like the appearance of the mortar a little better, although Jaki prefers the body and pointing quality of the sawdust mix.

Further tests on a 16-inch (40-centimeter) hardwood wall in Steuben, Wisconsin verified the results at Earthwood. Again, the Daratard 17 mortar was slightly superior, having fewer and smaller mortar shrinkage cracks. The sawdust we used at Steuben was rather fine, by the way, which probably worked against its ability to store moisture and retard the set.

An information sheet about Cemtex retarder, provided by Olle Lind, pointed toward a similar kind of retarder-to-cement ratio: "The normal rate of use is between 0.3 percent and 1.5 percent of the cement weight, with a maximum of 3.0 percent. An increase of 0.1 percent retards setting by an additional hour at 18 degrees Celsius (65 degrees Fahrenheit)." Similarly, Hans Hebel in Chile says of Sika Retarder: "It is a yellowish, milky liquid, and the normal mix is between 0.6 percent and 1 percent of the weight of the cement, but we put in about 1.6 percent to compensate also for the lime." This was a reasonable compensation, as Type S (hydrated or builder's) lime is considered to be a cementitious material. The reader will note that all three retarders tested so far carry a slightly different recommendation as to the amount to be used. However, the percentages are similar kinds of numbers; they do not vary by a factor of ten, for example. Note that all three companies (Cemtex, Grace, and Sika) express the amount of retarder in terms of cement weight. My advice is to test the brand you use, adjusting the exact percentage as you learn from earlier tests. This is what we did with Daratard 17, arriving finally at 3 ounces of retarder for 2 mounded shovelfuls (about 15 pounds) of Portland cement. Later, we found that the Sika Plastiment™ retarder can be used at about the same rate.

You can get a lot of 3-ounce batches from a gallon, so cement retarder is a good economy, as well as a good check on quality control. Also, just think of the labor saved by not hauling, screening, and soaking all that sawdust!

Making a Batch with Cement Retarder

Whichever brand of retarder you end up using, the method of introduction into the mix is the same. Follow these steps:

1. Using an ordinary garden hoe, mix your dry goods in the wheelbarrow until the mix has a consistent coloration throughout.
2. Make a little crater in the center of the dry mix.
3. Pour some water into the crater, perhaps half of what you think you'll need, making a little "crater lake."
4. Add the retarder to the water and mix it in.
5. Mix the mortar to the right consistency, adding extra water as needed. Use the snowball test described in Chapter 3.

A word of warning! A local cordwood builder discovered a major no-no, which cost him a day's work. In an effort to cover his bet, he made several batches of mortar with both soaked sawdust *and* Daratard 17. A couple of days later, he happened to walk along the top wall plate of the section he'd done this way, and the mortar crumbled beneath him. Normally, the mortar would have supported his weight easily. The hybrid sawdust-retarder mix had no structural integrity!

At Earthwood, we are very excited about the use of cement retarder (as a substitute for soaked sawdust) to slow the mortar set and, therefore, to reduce the incidence of mortar shrinkage cracks. Cement retarder can eliminate problems that sometimes occur when a builder is unable to get sawdust with the right characteristics. Experiment ... and share your results. The new Cordwood Builders Association (see Afterword) and Earthwood will maintain data on new developments.

CHAPTER 12

When It Shrinks, Stuff It!

Geoff Huggins

Humans create nothing that is free from drawbacks. Lest anyone think he or she has discovered utopia, just wait awhile, and the fly in the ointment will show up. I happen to believe that making cordwood walls is the ideal way to build a house. It is my utopian construction technique. But it does have its irritating fly: those gaps that inevitably open around the log-ends as the mortar dries.

For most of us cordwooders, this gap thing will happen; there's almost no way of escaping it. Why? To start with, mortar will not chemically adhere to wood, as it will to stones and bricks. Furthermore, appropriately dry wood laid up in a mortar wall will get wet, expand some, and then shrink as it dries again, pulling away from the mortar. Depending upon your type of wood, the initial moisture content of the wood, your construction technique, and other variables, you almost certainly will end up with gaps. The width of the gaps will range from miniscule to as much as a half-inch. If you are the fastidious type or don't want cold air or little critters coming in through these gaps, you may wish to stuff them.

Some Stuffing Requirements

Of the many stuffing options, go with the one that best appeals to you. I hope that some of the ideas in this section will help you in the selection process. I do not contend that I have the best answer or that I know more than anyone else on the subject. I've simply thought about the matter, have talked to others about it, and tried one idea way back in 1986 that has worked pretty well over the years.

What are some of the considerations to ponder when selecting a method to stuff those gaps? The material you choose and the technique you select will want to satisfy several criteria.

The ideal material should:

- Stick to both wood and mortar
- Be flexible (to allow for the seasonal changes in wood dimensions, as it grows and shrinks with moisture)
- Be economical
- Be easy to apply
- Be long lasting
- Look like mortar, so the patch is not visible or ugly.

I don't believe anyone has yet found the ideal material — at least to satisfy all of these criteria for all of us. So the choice is really one of selecting a material and technique that best fits your preferences. Using the preceding criteria list, I'll review a few of the options that have been suggested and used by others.

Latex caulking is a pretty good choice. It comes in handy tubes for application using one of those inexpensive caulking guns. It sticks to both wood and mortar. It is flexible, economical, and easy to apply. But latex does not look like mortar. It can be very noticeable and can even look, well, a little garish. Some cordwood builders have reported success in finding caulking which is intended to match gray mortar and has actually done so quite well. There is an element of luck in this, as everyone's mortar can be a slightly different color. But it's certainly worth trying a tube, particularly after your mortar has cured for a year, and has presumably achieved something like its final color. Remember, too, that it'll take a year of drying before your log-ends reach their final size. Clear caulking, one of Rob Roy's favorite methods, can be a reasonable choice, however. If your gaps are small, the clear caulking is virtually unnoticeable.

Silicone caulking is excellent but can be expensive. A much less expensive alternative is "siliconized" caulking, which has a lesser amount of silicone but still works extremely well. Red Devil Lifetime caulking and Cuprinol are just two brand names available. If you watch for sales, you can sometimes score these for two bucks or less per tube, whereas the full silicone formula will run from $4 (on sale) to $6 or $7 a tube — too much.

Caulking, including my own method described below, is a good stuffing choice if only a few — usually the larger — log-ends in the home shrink. If almost every log-end in the home shrinks, which will happen if you build with wood that has not had sufficient seasoning, you may want to go with a method that recoats the entire mortar joint.

Some folks have used Thoroseal™ (Thoro Corporation) or other masonry sealers. It's quick and economical, but the final product can look a little sloppy and long-term adherence can be a problem. And while it can stuff the gaps fairly effectively, it does not have the ability to expand and contract seasonally with the wood in the way that some of the other methods can. Thoroseal™ comes in 50-pound (23-kilogram) bags and is predominantly Portland cement with calcium stearate added as a waterproofing agent. It is normally used to waterproof concrete or concrete block foundations. Mix it to a thick paste and apply it with a knife or brush. Dampen the old mortar surface before the application (or paint on Acryl-60 bonding agent, also made by Thoro Corporation), but do not spray the whole cordwood wall, as you could cause a little wood expansion. In that case, when the wood shrinks again, you will have created new gaps. By the way, Thoroseal™ comes in white or gray. The white has the added benefit of brightening your home.

Perma Chink™ is a log cabin chinking product that has successfully been used by some, including cordwood writer and builder Richard Flatau, who is a strong advocate of this type of material. Perma Chink™ is made to look exactly like mortar, sticks well to both mortar and wood, is flexible, and long lasting. Also, for a variety of reasons, two adjacent batches of mortar in your wall may sometimes have different colors, so a uniformly colored coat of Perma Chink™ solves this problem. These materials are pricey, however. When we (Louisa, my house-building and house-inhabiting partner) built our house in 1985, Perma Chink™ had recently come on the market, and it was quite expensive. Log Jam™, made by Sashco Industries, is a similar kind of flexible chinking product. Rob Roy says it works extremely well, is very flexible, and comes in five different colors, including "mortar white." Unfortunately, it is even more expensive than Perma Chink™. Yet a third flexible log chinking product is made by Weatherseal.

You could compare these various log chinking products by purchasing a caulking tube of each — maybe the companies will send you a sample if you ask nicely — but when it comes time to do your project, buy in bulk (5-gallon [20-liter] buckets) not in caulking tubes, where the price per ounce begins to approximate that of illegal substances. Unlike caulking, these products are applied to the entire mortar joint. They will successfully close shrinkage gaps of a quarter-inch or more.

Incidentally, some cordwood builders have attempted to fill shrinkage gaps with aerosol-powered foam insulation, usually a garish orange in color. I have never seen or heard of a situation where this method has been anything less than, well, ugly. Similarly, oakum, a kind of a rope impregnated with oil-based goo, works very well to close gaps. So does fiberglass. But again, the appearance is sadly wanting and fiberglass can hold moisture. I wonder if a variation of Jim and Alan's papercrete or paper-enhanced mortar (Chapters 14 and 15), might make a good low-cost stuffing material that could be tuned to match the mortar.

These are some options. There are others, no doubt. Again, your choice can be sensibly made by weighing the criteria and selecting the material or method that fits your priorities. Let me now describe the one technique that I selected.

A Low-cost, But Not a Quick and Easy Choice

Louisa and I built our house on a very low budget. We had little money but plenty of time to spend on construction. In fact, we believe that a big advantage of cordwood — besides being beautiful — is that it is very inexpensive. Also, we were able to use wood from trees we cut right here on the land, so it was satisfying to use local material. But as many of you know — or have gathered by now — cordwood masonry is labor intensive. Since I'm a meticulous type (who wanted a neat-looking mortar face and clean, mortar-free log-ends when it was done), our cordwood walls required even more labor hours than most. So when it came to stuffing the inevitable gaps that occurred (despite my efforts to minimize them), a low-cost but labor-intensive method seemed appropriate.

I originally chose latex caulking because it's cheap, adheres to both wood and mortar, is flexible and long lasting, and easy to apply with a squeeze gun. While the latex is fresh, you can smooth and mold its surface. The main drawback of white latex, of course, is that it is quite visible, leaving a stark white ring around the log (see Image 12.1). So I mixed up a batch of dry mortar — from sand, lime, and cement, but not sawdust — in exactly the same proportions that I'd used in the walls. I then added lots of water to create a soupy, watery mix called a "slurry" of mortar. Using a small (say ½-inch paintbrush), I "painted" the slurry onto the white latex, blending it onto the surrounding mortar (see Image 12.2). It stuck to the face of the fresh, still gooey latex. When it had dried, the latex patch was quite invisible (see Image 12.3), looking as if the mortar had come right up to the logs. No gaps!

12.1: A small wedge beneath the log pushes it up so that the biggest gap is on the bottom, where it is less visible. Next, the gap is cleaned of loose mortar with any sharp tool, such as a nail (not shown). Particles and dust can be blown away with a small rubber hose. A bead of latex or acrylic caulking is applied and then smoothed with a popsicle stick, pointing knife, or finger. Credit: Geoff Huggins

12.2: The slurry or mortar is painted on and pressed into the caulk, then blended out to the surrounding mortar. Credit: Geoff Huggins

When I used this technique back in 1986, white latex caulking was all that was locally available. If you can purchase a clear or gray caulking as inexpensively as the white, those color choices would probably be easier to cover nicely with the mortar slurry. The white worked for us, but any system is open to improvement.

This is not a quick and easy method. I had lots of mortar painting to do. It helped me to avoid too much boredom by imagining that I was a sort of masonry Monet. But 16 years later, in 2002, the patches still look good and have held up very well. I demonstrated this technique on some of the large log-ends at the Pompanuck community round house during the 1999 Continental Cordwood Conference in Cambridge, New York. Within 20 minutes of application, the repair was practically seamless. My method was given rave reviews by all the cordwood gurus present, although I was probably pretty lucky with matching the color of the Pompanuck mortar.

12.3: Presto Farino! When dry, it is not easy to see the patch, particularly when the little wedge is removed.
Credit: Geoff Huggins

CHAPTER 13

A Mobile Home Converted to Cordwood

Al Fritsch and Jack Kieffer

Appalachia — Science in the Public Interest (ASPI), an appropriate technology center located in south central Kentucky, decided to cover the exterior walls of a 12-by-60-foot (3.6-by-18-meter) mobile home on the Community Land Trust property with cordwood made from donated scrap pine post cutoffs and "slabs" (the first cut taken off a log being ripped into timbers.) The property is located near Livingston, Kentucky, two miles from Exit 49 on Interstate 75. It is situated on a non-flooding bluff overlooking the Rockcastle River and is surrounded by the Daniel Boone National Forest.

13.1: Jack Kieffer's mobile home near Livingston, Kentucky was clad with cordwood masonry. Creidt: ASPI

Building Materials

The wood for the cordwood masonry siding was donated by the Kentucky Forest Products Company of London, Kentucky. Had this material not been used for the building, it would have been chipped as a wood processing by-product and sold to a local chip mill. A permanent roof of galvanized steel sheets, supported on pressure-treated 4-by-4-inch posts set in concrete, has covered the formerly leaking roof of the 24-year-old mobile home for about 14 years. A 2-foot (0.6-meter) overhang on both longer sides (north and south) allowed for the additional cordwood siding to be protected when added. The east and north walls are shown in Image 13.1.

The pine post cutoffs were cut into 12-inch (30-centimeter) log-ends and cemented with a Brixment™

mortar and sand mix on both interior and exterior ends, with an air space in the middle third of the wall. The space between the cordwood wall and the metal exterior wall of the mobile home was filled with half-inch Styrofoam™ insulation board.

General Features

Air vents were placed on the longer two sides of the building, along with an access door to the crawl space beneath the home. The foundation for the cordwood walls was made using 8-by-16-inch inch concrete blocks. The interior edge of the concrete blocks was placed on a concrete footer extending out from the mobile home skirting. The 12-inch log-ends were laid with a 2-inch (5-centimeter) overhang on each side of the block foundation. The foundation walls were plastered and flowerbeds installed on the sides where space permitted.

An added feature worth noting is that the mobile home is equipped with a cistern that collects rainwater and with a Carousel Dry Composting Toilet. The vault of the toilet is below the middle section of the home and has an access entrance on the south side (not seen in the photograph).

Door and Window Features

The doors and some of the windows for the home were donated by Habitat for Humanity. The windows came with complete framing, but the doors did not. The remaining windows were obtained from a construction materials recycling site at a cost of $2 each. The first set of windows (with the complete framing) were fitted with another frame made of two-by-sixes, in order to build up the depth of the frame.

The other windows were framed with a single surround of two-by-sixes. These windows hinge at the top so that they can be opened. The mobile home walls were disassembled and then reassembled to fit the dimensions of the new doors and windows. Windows framed with a single two-by-six surround were attached to the mobile home wall with lag bolts from the inside (see Image 13.2). These were long enough so that almost two inches (five centimeters) of the thread entered the frame, securing it to the mobile home wall. The cordwood wall was then built around the two-by-six frame as shown. In a similar manner, 2-by-12-inch door framing was attached to the mobile home wall after the wall was reconstructed to fit the doors. The cordwood wall was then built around the doors. Flashing was placed above the windows and doors to prevent leakage between the two walls. Usually the aluminum siding of the mobile home was attached to the top of the two-by-six (or two-by-twelve) framing to achieve this seal.

Wood Treatment

The exposed log-ends were treated with a one-to-one mix of turpentine and raw linseed oil, with a little paraffin added at the rate of one ounce per six cups of the mixture (30 milliliters per 1.4 liters). Writing in 2002, after several years of exposure, we feel that the wood treatment has been a success. Weathering is not too bad, and the wood seems very stable in the mortar matrix. The transpiration of moisture in and out of the log-ends is probably less than in an untreated wall.

Living in the Cordwood Building

There is a significant contrast between living in the mobile home before the addition of the cordwood walls and after they were built. The home is quieter, as road sounds are reduced. The double-paned windows are far more energy efficient than the original louvered windows. The cordwood outer wall has improved the heat retention during the cold times of the fall, winter, and spring. And the thermal mass of the cordwood has also helped to keep the home cool in the summer.

The thick walls provide wide windowsills for plants, which are attractive and help to filter the inside air. Since the roof of the mobile home is also rather thin, insulated panels with sheet metal sandwiching the insulation were placed on the roof of the mobile home. The new insulation helps to reduce heat loss through the roof.

People who pass by have stopped to look more carefully at the building and often inquire how it is made and how it is to live in. They like the looks of it and so do we.

We thank Mark Spencer for the drawing of the window detail.

For further information call Jack Kieffer at: (606) 453•3211.

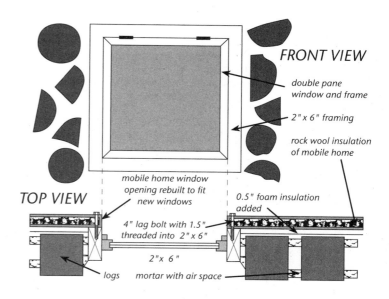

13.2: Window framing. Credit: Mark Spencer.

CHAPTER 14

A Shop Teacher's Approach

James S. Juczak

History

Taking the long view about human impact on our planet's ecosystems seems to be a product of getting older. The paths that my family have taken to be more Earth friendly have taken time but have come easily and naturally. There are a whole lot of things we are and are not. There are a whole lot of things we'd like to be. My wife Krista and I both feel somewhat trapped in the lifestyle we've chosen. We're both teachers; I teach "shop" and Krista teaches foreign languages. We're in our 40s and have two kids: Stacy, 15 and Elisabeth, an energetic 5.

My background includes teaching architectural drawing, construction, and related subjects since 1981. We presently live quite comfortably in a two-bedroom, ranch home that we've renovated over the past 16 years from a "wreck." Our five-year, mortgage-free, cordwood home project was a chance to "put my money where my mouth is."

Besides living in typical college accommodations, I've lived for a year-and-a-half in a converted school bus, and once lived out of my backpack during a year of hiking and hitchhiking. Krista has lived "camping style" for years, in such places such India, Peru, Germany, and Africa. Thus we both know a bit about how the other half lives.

The Heart of the Structure

After lengthy research (which included attending Earthwood Building School), lots of reading, and scoping out available recycled materials, we decided on an 18-sided, two-story design of about 42 feet (13 meters) in diameter. A central column would support the inner end of the radial floor joist and rafter systems.

The first part to be built was "the tower," made from precast concrete cylinders. I had been driving by a local industrial park — you know, "checking out the pickings" — and wondered what the local concrete products manufacturer did with the stuff that was made the wrong size for a particular job. A few phone calls and visits later I had my column. It worked out so easily that we couldn't believe how fast it went up. The general manager let me roam his "boneyard" and even helped me match my plans to what he had. We wound up with a series of manhole cylinders that gave us a 6-foot-diameter (1.8-meter-diameter) footed column with a concrete cap. Holes that were mistakes for him became openings for wood stoves, etc., for us. Another series of cylinders 4-feet-10-inches (1.5 meters) in diameter became the central column for our second floor. The 7-inch (18-centimeter) shelf, where the smaller column rests on the larger, supports the innermost end of our radial rafter system.

We wound up with a total column of 5,000-pound mix concrete reinforced with steel that is 23 feet (7 meters) tall. The $550 price included both delivery and set-up at my convenience. My dad and I poured the pad over a spring break, and the column raising took place in late June of 1999. The total assembly time for the column was under four hours. The column not only provides floor joists and rafter support, but also contains the masonry stove unit that will provide our heat.

Now (March 2002) the firebox and multiple flues are installed, the surplus openings are sealed, and the interior space of the heart of our home is about to be filled with sand. The central column will be an incredible heat sink, weighing in at over 30 tons (27 metric tonnes)! Once it is covered with an assortment of recycled tile, no one will suspect its humble beginnings.

14.1: The central supporting column, which would also become our mass stove, was erected in four hours. Credit: Jim Juczak.

We decided against a basement but didn't want a slab on grade. Both of us work on concrete and terrazzo floors and didn't want the same negative impact on our feet and backs at home. So a "ring beam" foundation was selected and placed on top of a base of compacted #2 crushed stone. This ring beam — sort of like a floating footer — is 12 inches (30 centimeters) thick and 32 inches (81 centimeters) wide. Its outer diameter is 44 feet (13.4 meters). This foundation gives us a 2-foot-high (0.6-meter-high) crawlspace under the home for running plumbing and wiring.

Tons of Framing Lumber

The post and beam frame of our now almost complete cordwood home is made out of recycled beams from a large bowling alley that was being demolished within six miles of our site. I asked the destruction foreman if I could get the wood from the 100-foot (30-meter) curved trusses that were being removed. I got 10 of the huge trusses, 400 sheets of used ⅝ inch roofing plywood, and about 500 pieces of 2-by-12-inch-by-21-foot framing lumber. Our cost was $10,000 for what I estimated to be over $50,000 worth of materials. Disassembling the trusses, denailing the lumber, and deroofing the plywood took me and a ragtag assortment of high school workers the better part of a summer to complete. The curved pieces became roof rafters; the straight laminated pieces became the 18 vertical posts in the outside wall; and the four-by material became the radial floor joists for the second floor. The first floor was radially framed with the 2-by-12-inch material and covered with two layers of recycled plywood.

We made several scale models to figure out how to use the material we'd scored. When my dad (a retired civil engineer) and I did the load calculations, we found out that the massive framing material eliminated the need for interior bearing posts between the central column and the external walls, freeing up the interior design quite a bit.

I took a shop teacher approach to assembling the frame. My students and I cut, drilled, and/or welded metal brackets for almost all of the lumber junctures. We first attempted to raise one of the 20-foot (6-meter) perimeter posts by hand; it was almost vertical when it decided to launch a crew of four off of the foundation in a most undignified and dramatic way. Luckily no one was injured, beyond a couple of bruises and scratches. Again, providence intervened. A family friend just happened to be dropping off a load of gravel that morning and had mentioned, just prior to the failed attempt, that he'd recently acquired a used logging bucket truck. After failing to "raise the flag" that day my dad commented, "Get the *%$# truck."

The posts and the primary joists for the second floor were raised in less than a day with a large crew of friends. Eighteen rafters, each weighing over 600 pounds (270 kilograms), took less than six (nerve-wracking) hours. I've skipped over a lot of the mindless preparatory work, but in less than ten days, we went from having just an 8-foot (2.4-meter) pad in the center of our future home to having the last primary rafter in place.

14.2: The logging bucket truck made post and rafter installation very much easier. Credit: Jim Juczak.

As a charter member of "Overbuilders Anonymous," I feel that the choice of these heavy timbers and many other really massively overbuilt parts were in keeping with our club philosophy. Communication, prefabricated steel brackets, the bucket truck, a nail gun, and an electric impact driver were key to the apparent ease of construction.

Hunt for Diamonds in the Rough

Dumpster hopping, garbage picking, and finding out what other people and businesses throw away has always intrigued me. The fine art of scrounging is of paramount importance in building a low-cost home. Never hesitate to ask about someone's apparent surplus! Here are some examples:

- Because a fire code changed, requiring steel doors to be installed in a local housing complex, I have all of the 1¾-inch wooden doors I'll ever need for the house — and plenty left over to barter with. Their cost? Five bucks each!
- The owner of several pieces of local rental property was going to destroy two cast iron tubs because they were too heavy to move back upstairs. He gave me the two tubs, two bathroom sinks, and a really neat toilet — just to save the cost of a dumpster.
- A local paper mill was experimenting with the production of automatic transmission gaskets, the base of which is a heavy polyester felt that's 6 feet (1.8 meters) wide. At zero cost, I got over 5,000 linear feet — 30,000 square feet! (2,800 square meters) — of this white, textured non-woven felt to use as interior wall covering and landscaping fabric.
- I called around to several manufacturers of replacement windows and found one that had 17 huge, low-E, argon-filled, vinyl, double-hung windows for $1,000 the lot.
- Hundreds of pieces of fixed insulated glass are available at a local glass company for less than half of their retail cost. They're typically called "orphans" because their owners just didn't pick them up after putting a hefty deposit down to have them made.
- I found a commercial six-burner cook stove for $300 from a local restaurant that had upgraded to a stainless steel model; and the list goes on and on ....

14.3: Our 16-inch-thick (40-centimeter-thick) walls have a solid papercrete mortar joint. The log-ends are pine. Credit: Jim Juczak.

Papercrete, or Paper-enhanced Mortar (PEM)

Our logs came from standing, dead red pine trees, which isn't so unusual. The mortar used in our cordwood masonry, however, is unusual, even by cordwood standards. The traditional cordwood masonry pattern is mortar, insulation, and mortar layers: we had vivid memories of Rob's M-I-M pattern stick that he uses as a teaching aid. I was frustrated with the slowness of M-I-M type masonry and the amount of fuss it took to get a smooth joint with a pointing knife. I also wanted to try to develop a material that more closely matched the insulation value of the log-ends themselves. I figured that a single homogeneous material laid up right through the wall would simplify construction. How could I get the insulation characteristics that would make this a viable option in our cold Northern New York climate?

The new mortar is made of paper sludge — 80 percent by volume — from the same mill where I got the felt. They throw away 40 cubic yards (30 cubic meters) of fiber-reinforced paper sludge every day! The other 20 percent of our "papercrete" is Type N masonry cement.

Incidentally, I agree that Alan's term, "paper-enhanced mortar" or "PEM," is more accurate, so will use it hereinafter. (See Chapter 15 for Alan's slightly different PEM experience.)

We mixed our PEM in 5-gallon (20-liter) buckets with a heavy-duty spackle blade on a half-inch drill, 100 gallons (380 liters) or so at a time. The PEM is both the structural support for the cordwood logs and the insulation at the same time.

Laying the 16-inch (40-centimeter) pine log-ends was a simple matter of dumping either a giant handful or even the entire bucketful of PEM and spreading the material across the foundation or the previous course of log-ends with rubber-gloved hands. We pointed the spaces between log-ends with gloved fingers first, then used a bent butter knife, and finally a stiff sponge to finish it off. The sponge absorbs quite a bit of the excess moisture from the papercrete and gives a slight stippling texture, which looks pretty good.

The next layer of log-ends are wiggled into place on the mortar bed, leaving about an inch (two-and-a-half centimeters) of space between pieces to facilitate pointing. The process is repeated — one bucket of mud after another, course after course. We found that a crew of five could lay up about 500 gallons (1,900 liters) of mortar on a good day, with two of those five mixing the PEM. This works out to about 150 square feet (14 square meters) of wall.

Typically, we would lay cordwood and mortar until just after lunch and spend the rest of the day pointing and tidying things up. One caution should

14.4 The cordwood is complete, summer 2001. Credit: Jim Juczak.

be observed with PEM: Don't lay cordwood masonry more than about 2 feet (0.6 meters) in height in a single day. The PEM is quite jelly-like, and the wall will start to lean in various directions if you try to build too high in a 24-hour period.

The paper-based mortar takes at least a day to set up and weeks to fully dry but has a hard finish similar to rigid foam insulation or hard papier-mâché. In the three years that I've been working with the stuff, there have been no cracks in the mortar, no settling of cordwood masonry within its panel, and minimal shrinkage gaps between the log-ends and the mortar. So the early returns on PEM are promising. Alan Stankevitz and Tom Huber, other cordwood building authors in this volume, have also been using various recipes of PEM with good results. Still, the reader is advised that PEM is a relatively new material, and all of the results are not in yet (such as its thermal performance).

Official Stuff

Getting the building permit was interesting. I waited quite a while to submit the proper paperwork. In fact, the entire frame was up by the time the code enforcement officer showed up for the first visit. I had documents ready for him, though: several sets of stamped plans (remember my dad the engineer), technical descriptions of the building process (which included photocopies of several pages of Rob Roy's book), copies of the site plan with distances from the roads and other property lines, and the results of several percolation tests at the leach field site. The papers included the math for both live- and dead-loading of the structure. A checkbook for the initial filing fee and a willingness to make changes in descriptions and plans to fit local requirements capped it all off without a hitch. (See Kris Dick's Chapter 27 for a more timely approach.)

Moving Time

When I told Krista we'd be into the new place in October of 2002, she replied that, "Yes, July would be great." I guess I need to move along a bit more quickly or plan on living with a camp atmosphere for a while. She is a bit eager to move. Right now the place is a habitable mess. Plumbing fixtures are in place, but nothing is connected. Most of the wiring is done, but only a few circuits are connected to the breaker box. There are no interior doors in place, and only a little over half of the drywall is installed. Nothing is spackled, the floors are unfinished, and then there's all of that dirt that has to somehow get onto the roof. But it's easy to heat with just one wood stove and, after all, it's home! *(Editor's Note: Jim, Krista, and the girls moved in on August 1, 2002.)*

CHAPTER 15

Paper-enhanced Mortar

Alan Stankevitz

Yes, the writing is in the wall. It is winter 2002. I have just finished my first year's work building a cordwood home in southeast Minnesota, using newspaper along with sand and Type N masonry cement. I call this mixture "paper-enhanced mortar" or "PEM" — a more accurate term, I think, than the term "papercrete," used rather casually with any cement and paper mixture.

Although the concept of using paper by-products in a cordwood mortar mix is still in its infancy, it is my opinion (rather than fact) that my current mix is "buildworthy." My experiment is being conducted in a cordwood wall that has a post and beam frame. My house is the two-story, 16-sided type described in Chapter 6, in combination with the Double Wall Technique discussed in Chapter 4. With load-bearing cordwood walls, I would be reluctant to use this mortar due to its lower strength.

I started my first wall, following in the footsteps of Jim Juczak — well, sort of. I've never seen Jim's shoes, and unfortunately, I am not close to a paper mill to get any free paper sludge. (Can you tell I'm jealous?) So instead, I was able to work out a deal with the county's recycling center for 75-pound (34-kilogram) bales of shredded newspaper. The first bales also contained office paper waste, which was hard to slurry. But after some "case of beer negotiations," the guys at the recycling center were happy to supply me with "pure" newspaper.

My first mix was a combination of two parts slurried newspaper to one part masonry cement. No sand. I'd tried this mix on a test wall with success, so it seemed okay to use on the house. The mix was very wet and hard to point; it had a slight cottage cheese texture to it. After I completed the first 8-foot square (2.4-meter square) panel of our home, I left it to dry for a couple of weeks ... then a couple more weeks ... and then a couple *more* weeks. While drying, the mortar color changed from dark gray to light gray, then to light green, and finally to a pleasing white. After six weeks, the PEM was pretty much dry. This wall was on the north side of the house, so the mortar probably would have dried faster elsewhere. Nonetheless, it was a slow process.

About one-third of the mix was type N masonry cement; the rest was slurried paper, made by soaking the bales of shredded paper for 24 hours in a 55-gallon (210-liter) drum. The mortar had no cracks in it, but I noticed a widening gap between the frame and the cordwood masonry. There were no gaps around individual log-ends, but the entire cordwood wall appeared to be shrinking.

Because of the shrinkage, I decided to try a more traditional cordwood mix, using sand with the paper and type N masonry cement. All subsequent walls have now been built using the following formula (by volume):

- 2 parts sand – 2 parts slurried paper – 1 part type N masonry cement

I love this mix. It has a very nice puttylike feel to it, dries in a couple of weeks, and looks great. Like Jim Juczak, I have been brush-finishing the mortar with a small foam painting brush, and when it dries you cannot detect any curing lines between batches.

When I built my test wall, I didn't lay cordwood inside of a frame like I did on the house. Had I done so, I would have discovered the shrinkage problem ahead of time and tried other mixes before building the real thing. The gap in the first panel, the one with no sand in the mortar, is now about a half-inch, and I can see daylight between the 6-by-6-inch post and beam frame and the wall. The 8-inch (20-centimeter) cordwood wall is still solid, as it is tied to the frame with frequent timber screws projecting two inches (five centimeters) into the solid mortar joint. Still, I'll probably rebuild the non-sand panel this spring (2002). I'd feel a lot better with a sand and paper mortar there.

Here are some pros and cons that I have observed:

Pros of PEM

- What a great way to use recycled newspaper! There's so much waste in the world — why not use it in an Earth friendly way?
- Paper doesn't have to cost you a penny. If you can't get it for free at a recycling center, just ask your friends, family, and neighbors to save it for you.

15.1: There are 16 panels on each of the two stories at the Stankevitz home. Built using a combination of Cliff Shockey's Double Wall Technique (see Chapter 4) and the Fraser framing method (see Chapter 6). Credit: Alan Stankevitz.

- The latest mix retards the set enough to eliminate most, if not all, mortar cracks but does not take forever to dry. All panels since the first have turned out fine.
- I don't really know the R-value of the PEM, but with its 40 percent paper content, I would assume that it's higher than more standard mortars. As per Cliff Shockey's Chapter 4, I will have insulation between my inner and outer cordwood walls.
- The PEM is visually appealing. You'd never know there is paper in the wall, and the mortar is a uniform, light gray color.

15.2: Paul Reavis built this cordwood shed with PEM mortar in Brodhead, Wisconisin, 1988. Credit: Alan Stankevitz

Cons of PEM

- Having to slurry the paper adds another step to the process. (If you are as lucky as Jim Juczak and have a friendly paper mill nearby, this extra step is eliminated.)
- I'm not sure how high up on a wall you could go in one day without the masonry compressing on itself a bit. If two people are laying, it might be wise to work on separate walls.
- PEM is not time tested yet. Only decades will tell how well the mix will hold up. I am encouraged, however, by a visit to a cordwood shed built by Paul Reavis of Brodhead, Wisconsin, in 1988. Paul's recipe started with the standard mason's mix of three parts sand to one part masonry cement. He bulked out the mix with unslurried paper, so that the final product was about 60 percent paper. Because the paper was not slurried, it did not become as integral a part of the mix. Nevertheless, the integrity of Paul's mortar is still good after 14 years.

Another PEM Formula by Tom Huber

Overall, my PEM experience has been similar to that of Alan's and Jim Juczak's, but my PEM recipe is a little different. I used various containers to measure out ingredients. Here it is by volume percentage. The whole numbers in parentheses are a close approximation of the mix in terms of proportion.

- 67 percent (17) shredded and soaked white office paper
- 21 percent (5) Type S "Mortar cement"
- 8 percent (2) mortar sand
- 4 percent (1) hydrated lime

The paper comes free from the college where I work. Type S Mortar cement is a masonry cement — not a "mortar mix" — which is masonry cement mixed with sand. Since I already had the lime to treat the sawdust insulation, I added a small amount to whiten the mix and give it a little more plasticity.

I initially "soften" the white shredded office paper by soaking it for a few days in a 55-gallon (210-liter) drum. I wring out the excess water from the paper and then mix it together with the dry ingredients in a wheelbarrow. I've used this mix M-I-M style — a 16-inch (40-centimeter) wall with cedar sawdust insulation — and as a solid mortar matrix laid transversely right through the wall.

We used PEM on the back wall of our "lodge" building (see Chapter 23). The only evidence of mortar shrinkage is a slight gap along a couple of posts, which I have caulked. I believe this gap was caused by the wicking action of the dry, rough-sawn pine posts, which draws the moisture out of the PEM quite quickly. I've also noticed a few thin tears or rips in the mortar after it completely dried in the M-I-M constructed wall. Again, the smaller volume of mortar in the M-I-M part probably dried faster. In regular cordwood mortar, such perforations might have manifested themselves as continuous cracks from log-end to log-end, but the reinforcing quality of the paper seems to stop continuous cracking.

I surmise that PEM concoctions of various kinds have good non-shrink characteristics, due to their moisture absorbent, slow-drying qualities. PEM has good "squishability" — more like bread dough than regular mortar — which makes it easy and pleasing to work with. It does take more time to mix, though.

15.3: Tom Huber's back wall is composed of 16-inch (40-centimeter) log-ends, PEM, and sawdust as insulation. The top section had not been pointed when the picture was taken. In the background is a small test building, also made with PEM. Credit: Tom Huber.

PEM: Observations from Rob Roy

My only experience with PEM came from putting down a few globs of the stuff at Jim Juczak's house during construction of his second panel. Jaki and I discovered that the stuff was quite pointable, and I think Jaki shamed Jim somewhat into pointing the subsequent panels by showing how attractive it could be. In the first panel, Jim just smoothened the PEM in a rough sort of way with his rubber gloves.

For this book, I have tried to keep to the "state of the art" subtitle by giving the latest PEM results from the new pioneers: Jim, Alan, and Tom. We are also encouraged by the success of walls built by Paul Reavis in 1988. None of us claims to be an expert on this technique, which must be considered to be still in its infancy.

Studying the writings submitted and personal interviews with the three primary PEM researchers have led me to the following observations:

1. The primary difference between the three "new pioneer" mixes is the sand content, which varied from none (Jim) to little (Tom) to a fair amount (Alan). Sand, obviously, makes the mortar harder and stronger but also denser. Sand increases thermal mass. The non-sand or low-sand mixes dry more slowly and would appear to have better insulation value.
2. The question of insulation value is still out with the jury. Obviously, PEM has a higher R-value than does regular cordwood mortar. How much higher is unknown, and the answer is one we all hope to have in time for the next cordwood conference. A related question is whether or not to go with a "solid" mortar joint through the wall or to employ the ordinary M-I-M style used with cordwood masonry. The performance of Jim's house through its first full north country winter in 2003 will go a long way toward answering that question. (You can email Jim at jsjuczak@mail.gisco.net)
3. Does PEM save time? Depends. With Jim's ready-mixed paper pulp and 45-second mixing time with a high-speed paddle drill, yes, mixing PEM is faster than mixing regular cordwood mortar. But both Alan and Tom report longer mixing times — Alan because of the extra time preparing the paper slurry, Tom because he hand mixes it like cob. Time spent on any mortar is largely a function of availability of materials.
4. Paper-enhanced mortar makes use of a waste material and shows promise as an insulative mortar that can be used in cordwood panels within a post and beam frame. It is too early to say whether or not PEM is suitable where the cordwood masonry is load-bearing. A test panel is advised with any new cordwood project (see Chapter 11), but is particularly advised in the case of PEM. And the test — or *tests*: try different mixes — should be conducted a full month before you want to start the actual project, due to the slower curing time.

Above: This stackwall barn, at least 150 years old, is on the French Settlement Road in South Gower Township, Grenville County, Ontario, near Kemptville. Credit: Wendy Huckabone.

Below: Earthwood. Credit: Rob Roy.

Cordwood Construction: The State of the Art

Right: Interior of Cliff's main house addition, built in 1990. Credit: Cliff Shockey.

Above: Cordwood interior by John and Edith Rylander in central Minnesota. Credit: John Rylander.

Right: Interior of round cordwood home on Michigan's Upper Peninsula. Credit: Rob Roy

Opposite: Clear bottles make the halo in the "Madonna" at Val and Jim Davidson's Marshwood house in Parson, British Columbia. The flower at child's level, Celtic cross and fish with bubbles are all at Marshwood and of Val's design. Credit: Jim and Val Davidson.

CORDWOOD CONSTRUCTION: THE STATE OF THE ART

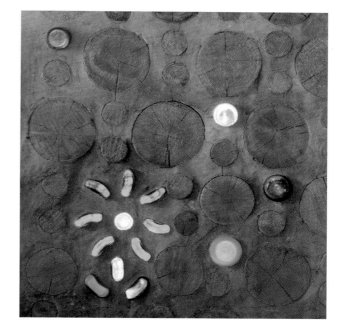
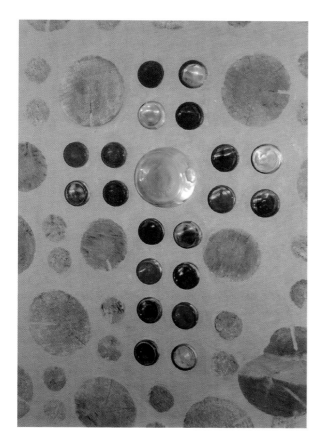
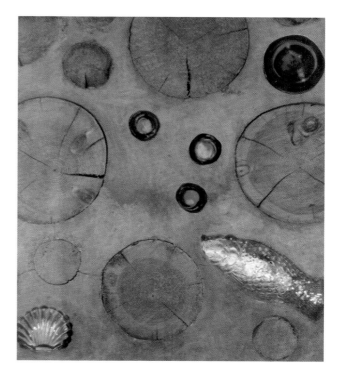

CORDWOOD CONSTRUCTION: THE STATE OF THE ART

Right: Stained glass panels fill the "snowblock" space between rafters at the cordwood home of Geoff Huggins and Louisa Poulin, near Winchester, Virginia. Geoff Huggins

Left: Wayne Higgins built his own circular stairs. Credit: Wayne Higgins.

Right: Interior of the den at Stonewood. Credit: Wayne Higgins.

Cordwood Construction: The State of the Art

Below: Stackwall cornered home in southern New York. Credit: Rob Roy

Above: Stonewood, the home of Wayne and Marlys Higgins in Calumet, Michigan. Credit: Wayne Higgins.

Below: "Woodland Treat," Hilton, New York Credit: Larry Schuth.

Above: Built by Steve Dunker, Cornucopia, Wisconsin, about 1969. Credit: Rob Roy

CORDWOOD CONSTRUCTION: THE STATE OF THE ART

*Right: The roundhouse at Brithdir Mawr, in western Wales.
Credit: Rob Roy*

*Below: Detail of Steen Moller's cobwood wall in Denmark.
Credit: Catherine Wanek.*

*Right center: Exterior wall detail at the home of Hans Hebel in Chile.
Credit: Hans Hebel.*

*Right: Pompanuck Farm Community Round House in Cambridge, New York.
Credit: John Carlson*

Above: One of two ventanas natureles at the Pompanuck community sauna. The little deva in the mortar protects the sauna from the Bannik, the evil sauna spirit.

Below: The lower entrance at Pompanuck.
Credits: John Carleson.

CORDWOOD CONSTRUCTION: THE STATE OF THE ART

Left: Our cat likes the circular stairway that leads to the loft bedroom. Right: Our dining room. Below: Our downstairs is largely open plan. Like most cordwood builders, we heat with wood. Credits: Ketter-McDiarmid Photos.

Part Three
The World of Cordwood Masonry

16 • Stonewood: A Love Story 117
   Wayne Higgins

17 • Woodland Treat 121
   Larry Schuth

18 • Cordwood on the Gulf Coast 127
   George Adkisson

19 • A Cordwood and Cob Roundhouse in Wales 133
   Tony Wrench

20 • More Cordwood and Cob 139
   Rob Roy

21 • Cordwood in Chile 147
   Hans Hebel

22 • One Old and One New in Sweden 151
   Olle Lind

23 • Creating with Stone, Wood, and Light 157
   Tom Huber

24 • The Community Round House at Pompanuck 167
   John Carlson and Scott Carrino

25 • A New Home on an Old Foundation 177
   Stephen and Christine Ketter-McDiarmid

CHAPTER 16

Stonewood: A Love Story

Wayne Higgins

It was a hot Saturday morning in August of 1968, the second day of a much-needed, three-day break from the stresses of work and the city. After a late morning breakfast in a small Ontario town, I was on the road again, in search of a log lodge said to be located on a wooded peninsula that jutted into Lake Huron's Georgian Bay. I'd received directions from a real estate office in town. The lodge was attractive in the picture and was modestly priced. I'd been interested in log structures for years and was eager to view this one.

I was soon traveling along a thin, twisting, macadam road that unsettled the stomach as it rose and fell. A white-hot sun climbed higher in a clear cerulean dome. Colors of objects in a shifting landscape appeared washed out as their shadows shortened. Tantalizing glimpses of Georgian Bay's cobalt-blue surface dazzled the senses. My destination must be near.

Suddenly, as I crested a rise, my foot stuttered forcefully on the brake, and I fishtailed onto the narrow shoulder. My eyes were locked to the right, and my pulse rate was climbing. "Oh, My, God!" was registered audibly. Seldom in my life had I been privileged to witness such beauty. She lounged sedately in dappled shade cast by a towering elm. Her graceful form and quiet presence lent an unimaginable dignity to an otherwise pedestrian grassy slope. While I struggled to regain my composure, I stared.

Did I dare approach her? Might I intrude, unannounced, upon her solitude and be forgiven? Quietly, I gathered my courage. With slow deliberation I left the car and scaled a wire fence, the only barrier between us.

I approached her obliquely, thinking that any direct route might be found unseemly or irreverent. Though she voiced no objection to my advances, she remained aloof. Did this unearthly ice maiden find me so wanting that she refused to acknowledge my presence? Then she spoke.

All that had suggested a cold demeanor melted in that instant. The veils were torn from my eyes. The floodgates opened, and the secrets of her inherent warmth and esthetic beauty flowed as a river. She spoke in languages at once foreign and familiar. She imparted to me her ancient lineage and her dreams for familial continuance. My heart swelled as I measured her with my eyes. I touched her. I caressed her. The desire to possess came over me.

Chance had driven me to her. There was an unspoken bond between us. Perhaps, somehow, I was meant to stay. Could I remain here with her? Was it remotely possible that I could take her with me? God, what was I thinking? I couldn't stay, and there was no argument on Earth that would move her. I reached out and made contact for the last time. Without saying goodbye, I turned and walked quickly down the sun-drenched slope to the road. A last, loving glance was cast as I drove away. Yes! Yes! Yes! I would find a way! One day she would be mine!

Twenty minutes later, I had located and offered an apology to the elderly Scots farmer who owned the land upon which — for the purpose of assignation — I'd trespassed. Nonplussed, the mild-mannered gentleman regaled me with the local history. He spoke of his family's role as early occupants of the land. He'd been preceded by five generations born and raised on this farm. He told also of the only other family who had, even earlier, settled the land. It was they who had brought the lovely creature into existence... over 200 years ago.

She — the object of my affection — was a small, white cedar cordwood barn. Petite, more aptly describes her. Certainly she was one of the most beautifully proportioned structures I've ever seen. She had a mature appearance, but she definitely did not look her age. Her gable ends were framed, and she supported a steeply pitched, galvanized metal roof with generous overhangs. Centered front and back were gabled dormers, similarly pitched. The gable-end windows were Gothic arched.

Initially, but mistakenly, I thought her to be a stone structure. She would have been strikingly beautiful even if that had been true. When she had finally revealed her nature, I was awed, deeply moved, and in love. I've since discovered that such beauty is a family trait, but until then I'd never before met a member of the cordwood family.

And the log lodge on Georgian Bay? I never made it.

THE PRECEDING STORY IS A TRUE ACCOUNT of the events that brought our Michigan home, Stonewood, into being. That weekend getaway led to my personal discovery of cordwood construction. The small structure spoke volumes, not only of the soundness of the method but also of its antiquity, its endurance, and its success.

This chapter's primary purpose isn't to address the how-to of cordwood masonry — others have done that — but to factually deal with why this traveler on life's highway was wakened to its possibilities.

That cordwood house in Ontario was one of the great loves of my life and served as the primary inspiration for our home near Calumet on Michigan's Upper Peninsula. I'll never forget her. Marlys, my wife of 30 years, remains my number one love. We have four adult children and three grandchildren, and all of them hold high rank on my list. Finally, there is Stonewood, our own cordwood home (built from 1989 to 1991), where we've lived for 11 years (see the Color Section). She is a work in progress and perhaps will always be so.

Our 2 acres (0.8 hectares) are situated on an ancient, hard-packed sand beach at Lake Superior's edge. I know it sounds improbable, but we never experience frost here once the thin layer of organic matter is scraped from the surface. Though I'd originally planned a rubble trench foundation, it wasn't necessary. I obtained our permit to build by submitting a floor plan on graph paper, a copy of Rob Roy's book, and a winning smile.

In the summer of 1989, I built an inch-to-the-foot scale model of the house to serve as the blueprint. Late that summer, we poured two strong concrete slabs and their footings. The footings are 14 inches (36 centimeters) deep by 20 inches (51 centimeters) wide. The slab is 6 inches (15 centimeters) deep from the footings toward the interior for a distance of 4 feet (1.2 meters), and then it tapers to 4 inches (10 centimeters). The cordwood wall thickness is 16 inches (40 centimeters, with 2 inches (5 centimeters) of the footing revealed outside. The concrete cured through a cold winter, with much of our cordwood stacked on the slab under tarps. We've only found one hairline crack in 12 years.

The following summer, in just under 90 days, my friend Wayne Remali and I built the cordwood walls to the top of the plates. The measured diagonals of the structure didn't vary by more than a quarter-inch. The door/window frames and lintels are 5-by-7-inch milled white cedar, doubled. We used Lomax units for our stackwall corners (see Chapter 5), also made from 5-by-7-inch stock. The 2-by-12-inch laminated rafters are hung from 5-by-14-inch ridge beams atop king and queen posts to ensure that the roof load is transferred vertically to the foundation. This structural detail permitted cathedral ceilings without the necessity of interior beams to tie the walls together. The roof was decked with ¾-inch plywood, upon which an 18-gauge, ribbed steel roof was attached. The floor system is a bit complex but results in a red oak surface 3 inches (8 centimeters) above the slab.

I designed and built our spiral stair (see the Color Section). It has fully cantilevered treads that carry the load. There are many other custom details that make the house unique. Examples include an antler chandelier, hand-carved lintels over the doors, and massive posts and beams incorporated into the structure. Our floor plan (upper and lower) contains 2,240 square feet (208 square meters). Marlys and I did all of the design work.

We had no negative experiences while building Stonewood. The logs are white cedar, taken from old barns that we dismantled on a nearby farm. They had been felled in 1905 and had remained perfectly sound.

Our mortar mix was found in Richard Flatau's book, *Cordwood Construction: A Log End View* (see the Bibliography). Richard's mix is: 3 parts sand, 3 parts soaked sawdust, 1 part

16.1: Stonewood, the home of Wayne and Marlys Higgins in Calumet, Michigan. Credit: Wayne Higgins.

Portland cement, and 1 part hydrated lime. There are very few cracks in the mortar and minimal shrinkage of wood. Cedar sawdust and lime make up the insulation. I've stuffed a little rolled or folded paper towel into a few checks from time to time, but we've never caulked.

A large cordwood project like Stonewood demands total commitment of its participants. Without my wife's commitment, particularly the financial support, it wouldn't have been possible. Thank you, Marlys!

Thanks also to "Cordwood Jack" Henstridge for his advice and great humor. He may recall the day that I phoned him and excitedly explained in some detail a new method I'd conceived for building corners. Finally he said, "Yeah, I know about it. We've been doing that for a while." Of course, it's called a Lomax corner.

To Rob Roy and Richard Flatau for their cordwood books, thanks. And also to Alan Stankevitz for his wonderful Daycreek website, where an earlier version of this chapter first appeared.

Finally, to my ol' friend and good neighbor, Wayne Remali … it's unlikely that I would have made it without you, eh!

I've spent my life as a designer, painter, illustrator, sculptor, and carver. We maintain a studio/showroom at Stonewood during the summer. Visitors — particularly cordwood enthusiasts — are always welcome, but appointments are recommended. The coffeepot is on 24/7/365. We live in Michigan's Upper Peninsula on the beautiful Keweenaw Peninsula. Write me at: 58091 Lakeshore Drive, Calumet, MI 58091; phone: (906) 337•9921; or email: whiggins@portup.com.

CHAPTER 17

Woodland Treat

Larry Schuth

The Beginning

IN 1973, AT THE AGE OF 33, I WAS GOING TO NIGHT SCHOOL, working my way up in a large company, raising a family, and working on a house. I was also reading *The Mother Earth News* and liked the idea of getting away from the rat race and "back to the land." To my wife Char and me, "back to the land" meant cordwood construction, which appealed because it had a very pleasing look and appeared to be both durable and doable. In 1982, I saw my first cordwood masonry in a chicken house in South Dakota that some students had built. It really looked great. I decided that if I ever had the chance to build cordwood, I would. In the late '80s, we borrowed an A-frame cabin in the Adirondacks and loved it. Char had always wanted a place in New York's Adirondacks, and I loved cordwood. We married these ideas and soon purchased 13 acres (5.3 hectares) of woods just outside the Adirondack Park. Here, we'd build our cabin. (See the Color Section.) We started to read everything about cordwood masonry that we could get our hands on. Two names kept popping up: Jack Henstridge and Rob Roy. We discovered that Rob conducted workshops not far away. So on a May weekend in 1993, we listened, looked at pictures, studied Rob and Jaki's house, got our hands dirty, asked questions, and were given a piece of cordwood to begin our own building — sort of an "Adam's rib" concept. We were off and running on our little timber framed cordwood cabin.

Our land was covered with 4- to 8-inch-diameter (10- to 20-centimeter-diameter) trees. It seemed like the stuff of log-end cabins. We were quick to learn, however, that maple, beech, ash, white birch, white pine, hop hornbeam, and yellow birch were not the most desirable species to build with. We knew from the workshop that white cedar was excellent — and there was some nearby — but not on our property.

Our first job was to select a building site and put in access from the dirt road. The property had a 30-foot-high (9-meter-high) ledge. There appeared to be granite bedrock near the top and a natural swale that led to the road. Using the swale to facilitate access seemed like a good idea, because there were not many trees in it and only a thin layer of soil there, over more bedrock. By adding gravel, we could fill in low spots and leave a drainage ditch on one side. But who could we find to build our road?

Gathering Wood

A bakery in town sold coffee to go with their sweet rolls. We were looking in their phone book for a contractor to build our road when a local lady asked if she could help. We explained what we were looking for, and she supplied us with names and phone numbers. In a small town, everybody knows everybody and knows what they do for a living. This can be a real asset. The first fellow we phoned built roads for a paper company, but his equipment was too big.

The next fellow asked about our project and gave us a price for roadwork. While we were considering his fee, he offered to throw in a truckload of cedar that he had at his place, about five miles (eight kilometers) away. That clinched the deal. He owned a large gravel pit, part of which was not used anymore and allowed us to use the space to cut, dry, and store our cedar until we were ready to use it. This was a great help, because our fully wooded property had no open sunny places in which to dry logs. His "truckload" turned out to be about nine face cords of 17-inch-long (43-centimeter-long) wood, our desired wall thickness. ("Face cords" 4 feet high, 8 feet long, and 17 inches wide are the measurements used in this chapter. With cordwood masonry, it is easier and less confusing to work in face cords than full cords.) Some of his cedar was over 20 inches (50 centimeters) in diameter, which we split. This fit perfectly into our plan to use a mix of round and split pieces in the walls. He also gave us about two cords of standing cedar and directed us to someone who had more that was already cut.

This second source of cedar was five miles (eight kilometers) away. This fellow had logged some of his woods and had the trunks cut into lumber and the treetops made into fence posts. But there is a section of each tree that is too small for timber and too large for fence posts. These sections were about 7 feet (2 meters) long and 6 to 9 inches (15 to 23 centimeters) in diameter. They had been stacked for about a year and the bark was falling off. We negotiated a price of a dollar each and bought 350 of them — about eight or nine cords.

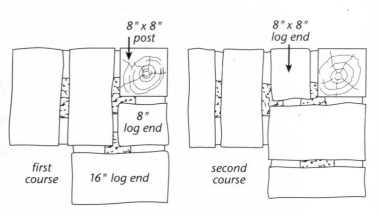

17.1: The use of a few short log-ends enables the Schulths to use a single 8-by-8-inch post in the corners.

I built a cutting table of old pipe, angle iron, an industrial door hinge, and a couple of rusty fence posts that would hold my 20-inch (50-centimeter) chainsaw at right angles to the working surface. There is a stop placed at seventeen inches. I placed marks on the cutting table at 17 inches (43 centimeters) and at 8½ inches (22 centimeters). If a piece was just over 17 inches, most of it fit on the table, and it was trimmed to size. If a piece was less than 17 inches, but more than 8½ inches, it was cut off at the shorter length and saved to dry. These half-length pieces are very useful in a timber frame building for placing opposite a timber or when building corners (see Image 17.1) We used so many of these that we actually had to cut a few full-sized ones in half.

We lived west of Rochester, New York, and I drove through suburbia on my way to work every day. At any time of the year people are disposing of cedar trees, because they are too big, broken because of ice or snow, or just not wanted any more. They leave them at the curb for pickup, and pick up I did. Over the next two years I collected about three cords this way.

An additional source of logs was found about two miles (three kilometers) from our cabin, where a cedar swamp was being cut during the winter by someone who was temporarily unemployed. Sometime in late winter, he gained employment and abandoned the swamp. In the spring, I noticed all the sawdust and scraps on the side of the road. I located the owner and asked if I could have the short scraps and explained what I was doing with them. He could not understand the concept of building with these scraps, so I invited him and his wife over to see, as we were well into laying logs by this time. The next time I saw him, he offered his trailer and all the abandoned logs that got lost during the winter's overnight snowfalls. All I had to do was drag them out to the road and load up the trailer. This yielded six more cords, none of which had to be split.

The last two cords were in one large cedar tree lying on the side of the road near a new ice cream store. The owner had cut it down so that people could see his sign. This tree cost $50, but the owner helped remove the bark.

Altogether, we cut about 31 face cords of 17-inch-long wood, and quite a few of the half-pieces. We ended up using 30 cords in our cabin. I have taken the time to tell of our cordwood acquisitions, because this is something most new builders worry about. Our experience shows that even white cedar can be available at low cost if eyes, ears, arms — and mouth — are kept open.

Building Permit

A sign at the edge of town proclaims "Building Permit Required." We set off to the county seat to inquired about getting one. I took my copy of Rob's *Complete Book of Cordwood*

Masonry Housebuilding (Sterling, 1992) and an idea of what we wanted for a structure. That's all. I introduced myself to two gentlemen at the building department and asked them what was needed to obtain a permit for a cordwood building. One inspector disappeared into his office, returning with a picture of the only cordwood house in his county. "Does it look like this?" I assured him that it did and opened the book.

We talked about buildings, timber sizes, snow loads, etc., for an hour-and-a-half. I drew a pencil sketch as we talked. It did not include any lumber sizes and allowed for the use of either logs or timbers to be used in building. I filled out a six-page form and 30 minutes later I had a permit in my hands.

You could have knocked me over with a feather! I had a permit, without furnishing architectural drawings, and that permit was valid for three years. No inspections of the footings were needed because we built on solid rock. No inspections before backfill, as there was very little. No inspections were needed for electric, heating or plumbing because we didn't have any. The whole process was much easier than we thought it would be.

Design and Frame

During the cordwood workshop at Earthwood, Rob suggested that we build a model of the house we had in mind. Using balsa, I built a scale model that didn't go together well. A trip to an architect was in order. He quickly provided solutions to my major concerns. He was so intrigued with the building concept, and pleased with the opportunity to finalize plans, that he rendered his services free of charge!

Our foundation was simple and inexpensive, thanks to building on bedrock. We simply cleaned the bedrock and built a 24-inch-high (60-centimeter-high) knee wall out of three courses of "doublewide" 8-inch (20-centimeter) concrete blocks. The inner and outer block walls, with a little mortar between them, gave us a 16-inch-wide (40-centimeter-wide) foundation. The log-ends are a half-inch proud of the knee wall, inside and out.

On the 4th of July 1994, three friends joined our family to erect the frame. I can stand up an 8-inch-by-8-inch-by-8-foot timber alone. Double the length, and it took four men, my wife, and ten-year-old son to stand each of eight of them in place. Cross members were tapped into place with a sledgehammer, and all joints were covered with a piece of 3-inch-wide (7.6-centimeter-wide) iron that allowed adjacent pieces to be lag-screwed together. Six-by-six-inch joists were added, and some temporary 1-by-10-inch floorboards were laid down and set with a few drywall screws for easy removal. Two weeks later, a couple friends came back to help put up the 4-by-6-inch roof rafters, 1-by-6-inch decking, 30-pound roofing felt, 4-inch (10-centimeter) foam insulation, and 1-by-4-inch purlins for the steel roofing to be screwed to. All sides of the building have a minimum of 2 feet (0.6 meters) of overhang for cordwood protection. (See the Color Section.)

Adding Character

Let your walls and logs speak to you. Initially we were unclear as to whether to build the walls with all rounds or all splits or a mixture. The mixture won out and we are happy. Another dilemma was how to add character to our walls. We discussed adding bottles, marbles, shells, and other items, but we finally decided to see what our woodpile had to offer.

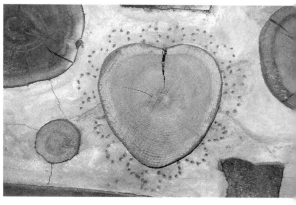

17.2: Cordwood heart. Credit: Larry Schuth.

Discovering interesting shapes in the end-grain lightened the task of cutting and splitting our wood. An angel, an apple, hearts, mushrooms, a butterfly, and an owl all showed up amongst our log-ends. Visiting friends have pointed out E.T. (the film star), a bowl of fruit, and at least 10 of the 50 states! A cordwood wall invites imagination.

Despite our conscious effort to minimize the amount of cement between the logs, we occasionally found large mortar areas, usually while pointing. We took advantage of these spots by carving flowers or patterns around the log-end. This added a little fun to the otherwise tedious job of pointing.

Not all pointing knives are created equal. We worked with a homemade steel knife, a couple of stainless steel tables knives, and silver-plated table knives. From all these options we quickly found our favorites —only two out of the whole bunch! Mine was a silver-plated knife; Char worked best with a thinner stainless steel knife. For reasons unknown to us, they were the only two knives that could bring up the gloss we were looking for. Char could not successfully point with mine nor I with hers. Progress ceased when one or both of us would misplace "our" knife.

17.3: Living room, Woodland Treat. Credit: Larry Schuth.

Not only did our family build the cabin of our dreams, but the cabin of our dreams built our family. While laying logs and pointing, we spent many hours in quiet conversation, sharing ideas, exchanging thoughts, and listening to the woods settle down in the evening as the thrushes, in song, bid the sun goodnight.

Schuth Cabin Statistics

Size:
- 900 square feet (83.6 square meters):
- 600 (55.7) on main floor,
- 300 (27.8) in sleeping loft

Type:
- Salt box design

Heat:
- Wood stove

Windows:
- 11 double hung, with brown anodized storms and screens
- 3 double hung with low-E glass, argon filled
- 1 casement with screen

Doors:
- Two, 4-inch-thick (10-centimeter-thick) solid cedar

Roof:
- Sheet metal, brown

Insulation:
- 5 cubic yards (3.8 cubic meters) of sawdust in the walls.
- 3 cubic yards (2.3 cubic meters) of sawdust in the cement
- 6-inch (15-centimeter) foam (some used, some scraps and waste) in floor.
- 4-inch (10-centimeter) Hy-therm roof insulation, R-28.4

Costs:
- Lumber $4,566
- Cement and Lime $1,160
- Insulation $960
- Roofing $865
- Windows $615
- Cedar logs $400
- Hardware $188
- Concrete blocks $154
- Sand $60
- Miscellaneous $302
- Total cost of cabin: $9,720 ($10.80 per square foot). Does not include grading, gravel, and heating.

CHAPTER 18

Cordwood on the Gulf Coast

George Adkisson

IN 1989 I REALIZED THAT I MUST DO SOMETHING to get my family out of the city lifestyle we were enduring. It didn't make sense to depend on services such as water, sewer, and garbage pickup, which are grossly overpriced through city taxes. Getting all our food from stores didn't seem right, either. I looked at several different building techniques that would allow us to make the break, and after weighing all the pros and cons of each, cordwood masonry seemed the obvious choice. We bought 3 acres (1.2 hectares) about 8 miles (13 kilometers) from the town of West Columbia, Texas, on the Gulf Coast.

I decided to keep the design simple: a two-story 30-by-40-foot (9-by-12-meter) rectangle with oak timbers every 8 feet (2.4 meters) along the side walls and every 6 feet (1.8 meters) on the gable ends. The upper story is entirely contained within a gambrel roof, with windows set into the steep sides of the gambrel. In the cordwood masonry lower story, doors and windows are framed out with 2-by-12-inch planks. Vertical sides of windows are framed from the wooden foundation plate (or "toe-plate") all the way to the sill plate (or "girt") that joins the tops of all the major oak posts. Horizontal boards were cut to size to complete the framing for each window. The 2-by-12-inch planks work in very well with our foot-thick cordwood walls, and framing the windows out ahead of time — and installing them — allowed us to move into the house before actually mortaring the cordwood into their panels. The foundation is a thickened-edge floating slab, 4 inches (10 centimeters) thick except for the 24-inch-thick (60-centimeter-thick) footings all around the perimeter and a 24 inch-thick (60-centimeter-thick) grade beam that runs down the middle of the 40-foot (12-meter) length. The grade beam supports the internal posts and walls under the center beam, which, in turn, support the upstairs floor joists and the whole center of the house. We used plenty of steel reinforcement in the foundation, as reinforcing rod (rebar) is much too cheap to be stingy with. We used four horizontal rows of rebar at different heights throughout the footings and grade beam, and heavy wire mesh throughout the slab. The 24-inch deep concrete perimeter footings and internal grade beam were placed just 6 inches (15 centimeters) below original grade, 18 inches (46 centimeters) above grade. The space beneath the slab floor was filled with compacted sand. The uninsulated slab has fared well and stays cool, even in the summer. I sprayed a termite treatment on the ground a foot or so out from the slab, and we walk around the building once in a while to see if the little varmints have built any of their mud tunnels up the outer edge of the slab. We have never been troubled by termites.

My choice of cordwood was easy. Our three acres was covered — and I do mean covered — with trees: mostly oaks, some elms, and a few red cedars.

We erected the home's framework, got the roof on, and then started stacking the green 12-inch (30-centimeter) log-ends between the heavy posts, without benefit of mortar. Because our log-ends were all hardwood, I decided to give them a full year of drying before I began to mortar them into the wall. So, we spent the winter of 1989–90 in the house, with just a half-inch layer of insulation board tacked up to the outer frame to keep out the elements. The thermometer hit 9 degrees Fahrenheit (-12.7 degrees Celsius) that winter, the coldest ever recorded down here.

We dried the wood one year and began the cordwood masonry in January of 1991. As it took us another year to finish the work, some of the last sections of the house walls actually benefited from two years of drying. Despite the long drying time, logs of 12 inches (30 centimeters) in diameter and greater shrank away from the mortar by as much as a quarter-inch (six millimeters). Logs of 6 inches (15 centimeters) in diameter or smaller do not show a gap, and I'd guess that, overall, probably one in ten log-ends are loose. This has not been a problem. I insulated between the inner and outer mortar joints with a 4-inch (10-centimeter) strip of ordinary fiberglass insulation, and this seems to prevent drafts from coming through the wall.

I've done nothing about filling the gaps but probably will someday. I might try Geoff Huggins's method (described in Chapter 12), which he demonstrated quite impressively at the 1999 Continental Cordwood Conference.

The mortar mix that I chose was the one Rob Roy used at Earthwood: 9 parts sand, 3 parts soaked sawdust, 3 parts masonry cement, 2 parts builder's (hydrated or Type S) lime. The sawdust was oak (all that was available), and I'm not sure it did any good other than to bulk out the mix. The ingredients were readily available and surprisingly cheap. Since the wall work was a slow process for one person, I bought the cement as I needed it — about $5 worth a week. The sand was mortar sand (washed and free of debris) and cost $125 for the whole house. The cost of the lime totaled about $60 and the sawdust was $25. By clearing the trees for the house and yard, the cost of the cordwood was lowered dramatically. The total cost of the outside wall of the ground floor came in at about $320!

By the time I was ready to build my shed and wellhouse a couple of years later, the good sand was used up, so I bought a pickup load of "fill" sand and screened it as I went along. It has done just as well as the washed sand.

Friends were convinced that the humidity would rot my untreated logs right down to the ground. Their fears were unfounded, and I believe that the key to the cordwood wall's continued excellent state of preservation is the "Texas porch" that surrounds the house. The logs have never been wet since the roof was up. In contrast, out of necessity, I had to have some quick but temporary shelter for the well equipment. I stacked the log-ends on the ground around the well and pump, with only a sheet of plywood nailed down at the structure's top corners to serve as a roof. The cordwood rotted in two years! I rebuilt the wellhouse properly, with plenty of overhang, and it has lasted. I see no need to apply any chemicals to the log-ends of the house. They are doing great by themselves, even in the near 100 percent humidity that we have all year long.

18.1: Part of our large "Texas porch."
Credit: George Adkisson.

The moist climate of the Texas Gulf Coast can actually be a big plus. I believe that the humidity — in combination with protection from the sun afforded by the large porch — retarded the curing of the mortar, minimizing mortar shrinkage and allowing me to point up to eight hours later.

The high humidity may also be the root cause of the slow but steady wood shrinkage. To address the possible shrinkage problems, I would suggest that no hardwood logs larger than 6 inches (15 centimeters) in diameter be used, and allow them a full two years to cure. This may seem like a terribly long time, but if you are living in the house while the wood is drying (as we did) and spend time working on the inside construction, it would be worth the wait. The protection of the Texas porch is important, too, in protecting the dry hardwood from the kind of expansion problems Rob speaks of (see Chapter 3).

To the best of my accounting, our 2,400-square-foot (223-square-meter) house cost about $25,000, or just over $10 per square foot. This includes a $4,800 forced-air, wood-burning fireplace that I had installed. If you find it beyond your abilities to build a masonry stove, I highly recommend this type of wood-burning heater. We also carpeted the entire house, with the exception of the kitchen, which has a vinyl floor covering. Two large, window-mounted air conditioners keep us comfortable in the hot months from May to September — and with a fraction of the electricity that neighbors use.

The $25,000 cost estimate is very unevenly divided. The upstairs portion, with its 2-by-6-inch standard frame construction, is where most of the money went. If I were to do it again, I would probably build a 60-by-80-foot (18-by-24-meter) single-story home, and get twice the area for about the same money, or even less.

To finance my building adventure, I cashed in a retirement fund early. This was done before the government enacted penalties on this sort of thing. Thankfully, I didn't have to seek financing, as this would have been impossible back then. To secure financing where we live, one must purchase hurricane insurance. I petitioned the insurance board to accept my structure for insurance, but they replied that they were not sure if this construction style would withstand hurricane winds and that they were not about to research the question for one potential customer. However, an open-minded insurance inspector gave me good information on how to make the building hurricane-proof. For example, I fastened the toe-plate to the slab with anchor bolts, and used special right-angle metal connectors to fasten the wall posts to both the toe-plate and the plate beam. I built it strong enough (I hope) to withstand any winds. About five years later, the insurance board redrew the geographical lines where they believed homes would be affected by a hurricane, and we now have hurricane insurance for a very small premium.

The only other code issue I had to deal with was the septic system, which was plainly illustrated in my application and not affected by the style of the house.

During the time we lived in the house with the log-ends stacked without mortar, bugs became a serious concern. Wood grubs are insect larvae laid in the bark, which I had not yet removed from my log-ends. At night, with the TV off, I could hear them chomping away. This seemed like a problem that would be impossible to deal with. By spring of the first year, the mature bugs were beginning to come out. I collected a few and went to see the county agriculture agent. He listened to my description of the house and promptly said, "Live with it for two years and then they'll be gone. Barking the wood will break their life cycle. They won't come back." And he was right.

My kids, now adults, can't believe how beautiful the house turned out. They also swear they'll never debark another log as long as they live and say it is too hard to build your own house. But I expect that one day they will grow weary of working for next to nothing, and they'll be back looking for a little help. Dad'll have it figured out, I can hear them saying. And thanks to some professional advice and help along the way, I guess I do.

Life BC (before cordwood) was very different from today. My family and I are no longer slaves to house payments, high utility bills, city life, or cramped living conditions. To rise each morning at dawn and watch the white-tailed deer and squirrels literally playing in the backyard is powerful medicine for the heart and soul. To back up to a warm wood fire that quickly heats the whole house on a cold morning brings the family together for warmth, conversation, and laughter. This fire is a very popular spot from November through February. The garden supplies us with most of our vegetables for the year. An untainted supply of food benefits body and soul. The combination of our new lifestyle with some modern conveniences — a washer and dryer, microwave, TV, and computer — has improved the quality of our lives many times over.

I owe a debt of gratitude to all of the new pioneer cordwood builders who helped resurrect this profoundly simple, yet strong and beautiful building technique. My lack of structural knowledge was easily overcome by the power of foot-thick walls. And who knows? Maybe it has been some small contribution for us to show that cordwood masonry can even work under the humid Gulf Coast conditions of Texas, as long as basic precautions are taken.

Epilogue, February 2002

The foregoing was written for the 1999 Cordwood Conference. In May of 2000, Gwen and I hosted a three-day cordwood workshop with Rob and Jaki Roy. We had 18 students at our home to learn the finer points of cordwood building. I realized then that a workshop would have taught me techniques that would have saved me dollars and time.

The workshop project was a simple eight-sided gazebo. Two of the sides are completely infilled with cordwood (and small windows), in order to provide rigidity to the structure. The other sides are cordwood only one-quarter of the way up. The eight posts are made of two, 2-by-10 posts set at a 135-degree angle to each other, with the gap filled with small log-ends and a 2-by-4-inch spacer on the outside only.

The large open windows afford us a total view of the house, gardens, and yard. We enjoy the gazebo in the mornings with coffee, before the daily activities begin, and the sun gleams through the colored bottle-ends on the east wall. We often return in the evenings, after working in the garden. And it makes a great playhouse for the grandkids.

Although our home has not yet been tested by a hurricane, in 2001 it was severely tested by rain. Had

18.2: This gazebo was built at the Adkisson property during a workshop conducted by Rob and Jaki Roy.
Credit: George Adkisson.

we received just 1.75 inches more rain, it would have been the wettest year ever recorded here. But, at 71.18 inches, it seemed like a Houston area record to me. We have not experienced any problems with any of our four cordwood structures in the Gulf Coast climate.

To sum up, I can only say, find a way to build it, and your life will change forever.

CHAPTER 19

A Cordwood and Cob Roundhouse in Wales

Tony Wrench

Our round earth-sheltered cordwood and cob house is sited at Brithdir Mawr in West Wales, a community aiming at sustainability. (See the Color Section.) The Roundhouse is on the edge of deciduous woods. Our objectives with this building were that it be autonomous and have very low environmental impact and low embodied energy. Therefore, we used no cement, no sawn timber, and no new glass.

The diagrams by Olwyn (a friend from West Wales), I hope, give you a pretty good idea of our construction methods. Here are some other facts that may be of interest:

1. The timber is Douglas fir, from 1.5 acres (0.6 hectares) of forestry that Jane, (my partner) and I bought in 1995. I designed the house in general terms in the autumn of 1996 and spent that winter cutting 200 of the smaller trees by hand. Some of these were 20 to 23 feet long (6 to 7 meters long), for use as primary roof timbers. The majority were 13 feet long (4 meters long), for use as poles, secondary roof timbers, and cordwood. The forest is still there, of course, as the trees I cut were only thinnings. The wood was transported 20 miles (32 kilometers) to our community in two truckloads. (There is no fir at Brithdir Mawr, only deciduous trees.) The material was carried the final distance to our site by tractor and trailer.

   With an electric chainsaw, I cut about 90 percent of the logs near the main house of the community, where we have 12-volt power plus an inverter, and mains (commercial electric) at 240 volts. Again, the 16-inch (40-centimeter) log-ends were delivered to site by tractor and trailer. We used no power tools on the site, so the house was built using hand tools only.

2. The structure's skeleton is poles. Cordwood is used as infill and is not intended to be load-bearing. I wanted a heavy turf roof, so the design includes plenty of poles and plenty of rafters, as can be seen in the diagrams. I chose 16-inch (40-centimeter) log-ends because I had read in an American book on log cabins that heat travels 2½ times faster along the grain than across it, so I reckoned that a 16-inch cordwood wall would be equal to a 6.4-inch-thick (16-centimeter-thick) solid wall.

3. The clay, which we used in place of cement mortar, was all taken from site. A JCB (backhoe) dug a circle into the bank, and the subsoil turned out to be clay with a bit of sand. We just mixed this with rainwater and a couple of handfuls of straw per wheelbarrow load and wound up with a very good cob material, as has been used for construction in Britain for over 1,000 years. We used a bit of bracken in the cob, as well, but not much. There has been some shrinkage around the log-ends. Gaps of up to a half-inch (one centimeter) have appeared, especially around the south side, but these are easy to fill with a little additional cob.

4. At the eaves, the 120 radial rafters cross the wall plate about 9 inches (23 centimeters) apart. The gaps between rafters — what you call "snow-blocking" in North America — are filled with logs about 10 inches (25 centimeters) long and then plugged with a 50/50 hay-and-clay mix rolled into clumps or balls. I thought this might be better insulation. The 50/50 mix worked well for filling larger holes and it set very hard. Plus there is nice potential along the eaves for small birds to find good nesting sites.

5. The floor is just beaten clay, a bit bumpy, but fine with carpets on it. The floor is probably our main source of heat loss, but the house has been warm enough that the heat loss through the floor hasn't bothered us.

6. The stove surround and the warm internal wall containing the flue are also made of clay. We used a slightly different mix for them: approximately 80 percent fine gray clay dug from a nearby lake and 20 percent fine silt from the lake outlet. This mix has almost no organic content and has set really hard around the metal of the stove.

7. By using this style of construction, I found the building process to be delightful as well as natural. The cordwood process is very similar to dry stone wall building, in that you have to choose each log individually from a big pile. We used branch stubs and other protrusions as anchoring into the cob. Big logs of 12 inches (30 centimeters) or so in diameter are lovely to use, and one makes good progress with them. It was a bit of a fiddle fitting logs around the diagonal bracing, because I used 9-inch-long (23-centimeter-long) logs up to the diagonal. But these still needed to provide good bearing for the full-length 12-inch logs above the diagonal. In the six wall panels with a diagonal, I found I needed a lot of 2- to 3-inch-diameter (5- to 8-

centimeter-diameter) logs to fill the smaller spaces. Some of these spaces could also be filled by a wine bottle plugged into a jam jar.

8. Each of the 13 wall sections took about a week to build. Windows were large and rather fiddly to fit accurately. Straw bales, used in a few places where they would not be underground or exposed to driving rain, saved a lot of logs and were easy to plaster with the clay mix.

9. I liked the absence of building waste on this project. Generally, there is nothing about the building that I would change. It simply feels good! If, some day, we want a fireplace with opening doors so that we can watch an open fire, we will probably need to widen the chimney flue or perhaps increase its height. As it stands now, the warm wall containing the sloping flue works very well, and the stove stays on most nights. The outside solar water heating panel is great. A whiskey barrel makes a good hot water tank, though you should steam-clean the inside before using it. Well, that's about it, folks. Enjoy Olwyn's fine captioned drawings.

(Editor's Note: Since writing the preceding story for the 1999 Cordwood Conference Papers, Tony has added a Villager wood stove with a back boiler and raised the chimney 6 feet, 6 inches (2 meters) to improve the draw. He has also installed a wood floor of sawn larch planks in the central area, covered with handmade wool rugs made on a peg loom. The house is certainly homey and comfy, as it was when I visited in September of 1999.

As I write (December 2002), Tony's house is under threat by the planning authorities, who have ordered him to tear it down by March of 2003. This battle has been going on for years and would be laughable if it weren't so serious. While saying that development should be "sustainable," the planners fail to recognize that Tony's house is one of the most sustainable in all of Britain. You can get the whole story on Tony's website at: www.thatroundhouse.info. Also, Tony has written a fine little book (see the Bibliography) about his roundhouse.)

19.1: The Roundhouse at Brithdir Mawr, in western Wales. Credit: Olwyn

LOW IMPACT TURF ROOFED ROUND HOUSE

An experimental structure built into a south facing bank at the head of a small valley.

The bracken covered slope was dug out using a J.C.B. and the building set into the semi-circular excavation. Visually it blends in well — It is nearly invisible from a distance. Materials: Soft-wood plantation thinnings, recycled windows, most of which are double glazed, clay, and some rubber, (N.B. Shingles on front - made from old tractor inner tubes). It is designed to last 30-40 years and then biodegrade!

Structure by Tony Wrench
Article by Olwyn

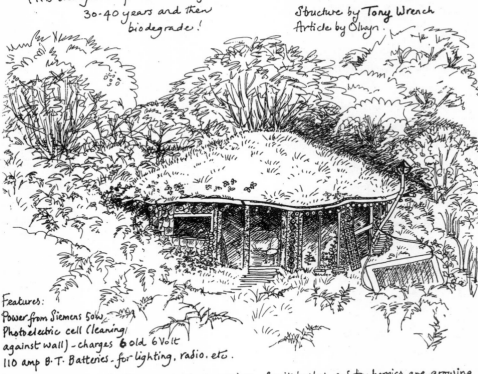

Features:
Power from Siemens 50w Photoelectric cell (leaning against wall) - charges 6 old 6 Volt 110 amp B.T. Batteries - for lighting, radio. etc.

Strawberries in roof & by front wall - 4 grapevines, fruit bushes and tayberries are growing on or round the structure. Reed bed on right hand side for waste water recycling. Bottles with jars on the end, set into wall, create a stained glass effect from inside.

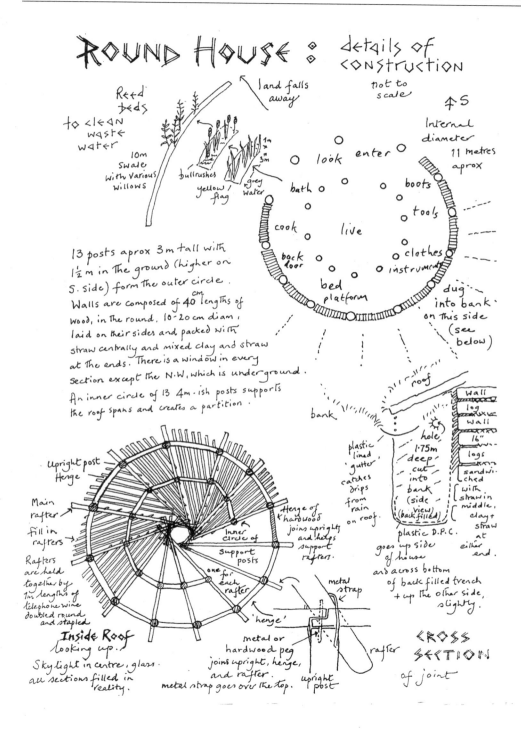

19.2: The Roundhouse at Brithdir Mawr, in western Wales. Credit: Olwyn

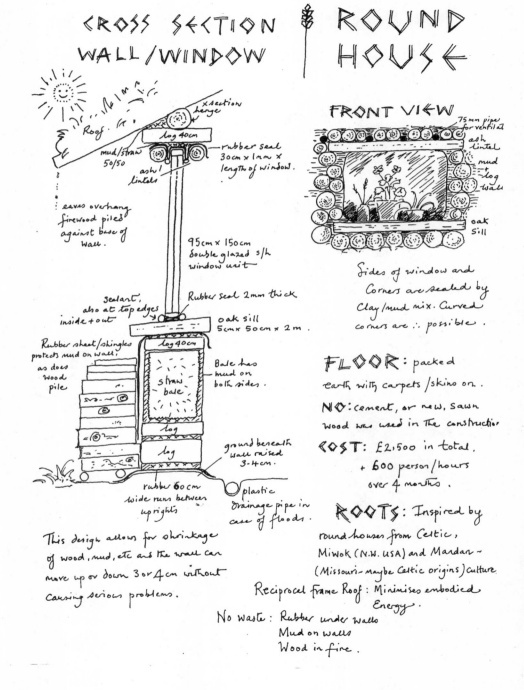

19.3: The Roundhouse at Brithdir Mawr, in western Wales. Credit: Olwyn.

CHAPTER 20

More Cordwood and Cob

Rob Roy

In Chapter 2, I admitted to a 25-year love affair with cordwood masonry. I love it so much that I accept its warts, primarily its labor intensiveness and the vagaries of wood expansion and shrinkage. Mortar shrinkage has not been a problem for us since we discovered the soaked sawdust additive many years ago. But more recently, there has been a greater awareness about the energy cost of materials. Just as fuel consumption should be an important consideration in deciding on a vehicle purchase, the "embodied energy" of building materials (the amount of energy required to manufacture those materials) should be factored in at the design stage of home building. This is one of the major points made by the "natural building" movement of the past five years or so. And it's a good point, as it takes 500,000 BTUs of energy to make a bag of Portland cement — equivalent to 4 gallons (15 liters) of gasoline. And there are other environmental considerations, such as air quality impact and the fact that limestone quarries lay waste to a lot of land. Is there a natural material that can substitute for the mortar in a cordwood wall? I believe there is, and it's called "cob."

"Cob" in Old English, means a lump or rounded mass. In construction, it has come to refer to building with earth, usually one "cob" at a time — each one the size of a small loaf. At least one-third of the world's people live in earthen homes of one kind or another (cob, adobe, rammed earth, etc.) and have been doing so since the beginning of human history. Many cob houses in the southwest of England date back 500 years, and some are twice that old. Author Dan Chiras, in *The Natural House* (Chelsea Green, 2000) (see the Bibliography), points out that swallows build little "cob cottages" one beakful at a time and wonders if humans got the idea from our feathered friends.

Ianto Evans and Linda Smiley have taken cob to a higher, somewhat more scientific standard, although they are still careful to follow traditional cob building techniques where these have passed the test of time. At their North American School of Natural Building in Oregon (www.deatech.com/cobcottage/), they teach the use of "Oregon Cob," with ingredients falling within certain proven parameters to assure maximum strength. More on this later.

Now cordwood masonry is time consuming, but so is making a wall of solid cob. What if the techniques were combined? Cutting the labor-intensive cob-mixing process down to, say, 30 to 40 percent of that used in a solid cob wall has a compelling attraction. But can it be done? Well obviously, yes it can, as Tony Wrench has demonstrated in his Roundhouse (see Chapter 19). Instead of mixing mortar, Tony simply mixed the earth at his feet with rainwater and "a couple of handfuls of straw per wheelbarrow load," and wound up with "a very good cob material, as has been used for construction in Britain for over 1,000 years." Tony's soil, he thought, was high in clay "with a bit of sand." This cob material was used in place of mortar, and the insulated space between the inner and outer cob joints was filled with straw for insulation.

I visited Tony's cordwood and cob Roundhouse in 1999, the first time I'd seen the combination, which has now come to be known as "cobwood". As infilling, the system was obviously sound, thought the surface texture could have been improved with pointing. (A year later, Jaki and I found that cob can be pointed quite nicely.) When Ianto Evans made a journey back to his native Wales during the winter of 2002, he stopped in at Tony's place and was impressed with the overall package, although he told me that Tony's soil was actually quite high in silt more than clay — the two can be easily confused, even by experts — which tends to yield a cob of less hardness and structural ability. As infilling, though, Ianto saw no problem with Tony's mix.

Cobwood at Earthwood

Jaki and I had just finished hosting a major, four-day Megalithics (stone circle) workshop at Earthwood, and our energies were still high when Linda and Ianto arrived, fresh from a cob workshop they had conducted in Ontario. We all had an interest in trying cobwood construction and figured that the four of us ought to be able to work it out, if anyone could.

The garage had been recently completed with standard cordwood mortar throughout. The frame was 8-by-8-inch timbers, establishing an 8-inch-thick (20-centimeter-thick) cordwood wall. The log-ends were mostly dry spruce, three years cut and dried, with a few pieces of white cedar, poplar, and basswood, also very dry. The wall panels were all protected by at least 2 feet (0.6 meters) of overhang. Jaki and I decided to remove a translucent 4-foot-by-6-foot-8-inch sheet of Filon fiberglass greenhouse covering from one of the panels and replace it with cobwood. We figured we could get by with a little less light in the building, in the interests of science.

Our first problem was that we had not a speck of clay on top of the hill where we live, and Ianto can literally smell the stuff out. I called a contractor friend and learned that he'd done an excavation near Lake Champlain, 15 miles (24 kilometers) away, and that he didn't think the owners would mind if we took a little of the messy, greasy stuff. Ianto and I took my pickup truck over to the site and loaded on enough sticky, gray clay to make cob for the test panel. When we got back to Earthwood, Ianto began soaking the clay clumps in 5-gallon (20-liter) buckets, to hydrate (or soften) it so that it would be ready for use the next day.

Another missing ingredient was good straw. We had some rotting straw, but Ianto said that wouldn't do, so we ended up using some dry, fairly coarse hay instead of straw. (Straw is preferable, though, because its high cellulose content prevents it from breaking down easily.) Fortunately, we did have plenty of sand on site — coarse and fine.

The cob experts could tell by feel that the clay we'd found was quite pure, so they recommended a mix that would be about 20 percent clay and 80 percent sand by volume. Some builders may be fortunate in having earth on hand with a favorable combination of clay and sand. Generally, an earth with clay content of from 10 to 30 percent will yield pretty good cob.

But how can you tell?

Cobbers recommend the "shake test." First, pulverize a sample of your soil and put some of it in a clear quart Mason's jar, say one-third of the jar. Then add water so that the jar is almost full. An added teaspoon of salt is said to help any clay to settle out. Put the lid on and shake the heck out of the mix. Really shake it! Then, set it level and watch the earth materials settle out. Within 3 seconds, the pebbles and the coarse sand will settle to the bottom. Silt or very fine sand may take 10 minutes to settle out. If there is no clay in the soil, the water will be fairly clear at this point. In this case, clay would need to be imported to mix with your soil for cob-making. If clay is present, it'll be suspended in the water after 10 or 15 minutes. It may take hours to days for the clay to settle out, but once the water is clear, you can determine the percentages of sand and silt. You will also know if there is clay, but Ianto tells me that the shake test will not give you a very precise percentage of clay unless the water is somehow removed — a very difficult process.

Shovelfuls of our sand and hydrated clay were piled in the middle of a 6-by-8-foot (1.8-by-2.4-meter) blue polyvinyl tarp lying on the middle of the garage floor. Best to put the sand down first, as the clay tends to stick to the tarp. The ingredients were added at the rate of four parts sand to one part of our fairly pure hydrated clay. After a manageable amount was assembled in the middle of the tarp — about a 5-gallon (20-liter) pail full — Linda taught us to turn the ingredients by lifting the edges of the tarp, always folding the goods into the center. This goes better with two people, one on each side of the tarp.

Turn the goods until the clay clumps are broken up and the mix has taken on a fairly consistent coloration.

After turning, the mix is danced on by the cobbers. Jaki and Linda each wore rubber wading shoes, although many cobbers with toughened feet perform this operation barefooted. Sharp particles, however, can cut tender feet, so be warned. Know your material and your feet. Personally, I prefer the protection of the rubber wading shoes. I was reminded of the romantic vision of crushing grapes at a winery. The purpose of this dance, however, is to drive the tiny clay platelets into the voids between the sand grains. Sand gives the cob its hardness and non-shrink characteristics, while the clay acts as the cement that bonds the material together and gives it strength. The clay can be thought of as a natural cement when used in this way.

Water can be added to give the cob a good plastic consistency and texture. Then, straw (in our test, coarse hay) can be shaken into the mix from the flakes of bales and pressed in with your feet. More straw and water can be added as needed, and you will find it handy to turn the mix over now and again by lifting corners of the tarp. How much straw? After a while, the cob will feel like a tough cohesive mixture, as opposed to squishy mud.

We started out making cob in pretty much the same way as Ianto and Linda would prepare it for a solid cob wall. We had to start somewhere.

Then it was time to experiment with the cobwood technique. Jaki and I laid out the cob on the 8-inch-wide (20-centimeter-wide) wooden base of the panel as if it were mortar. We found it difficult to work with because of the long fibers of hay. We experimented with the M-I-M method (except that we were using cob instead of mortar) but also tested a solid cob joint transversely through the wall. We showed Ianto and Linda how we set the log-ends and found that it was much the same as with mortar — except the cob was stiffer than ordinary cordwood mud. The tough part was pressing the long-fibered cob into the spaces between log-ends. It worked well to use long sausagelike cobs with a cross-section similar to the mortar for which we were substituting.

On the next batch, we tried shorter hay fibers, made by running our rotary lawn mower over a section of the hay bale. In no time, we had fibers two to three inches (five to eight centimeters) long. This cob was easier to mix and much easier to lay between the log-ends.

20.1: Linda Smiley (left) and Jaki Roy lay up cob in preparation for log-ends. Note that rubber gloves are not necessary for playing with "mud pies." Panel at right is cordwood with mortar.

There was some grumbling amongst us, as we each wanted to apply our long-standing methods of working with our pet materials to the project. But we also had a lot of fun, and after a day's work we had learned quite a bit. Primarily, although we were in the first hours of the test, we were all optimistic that it was going to be a success. Ianto and Linda were happy with the way their cob was performing, and our log-ends didn't seem to mind being laid up with cob instead of mortar. In fact, the wall took on the appearance of a more or less ordinary cordwood wall, except that the "mortar" was brown rather than gray.

Linda also showed us how to make a finer cob for plastering or pointing. This mix made use of much finer sand that we had on hand, a higher percentage of sand, finely chopped straw, and more water. This plaster-grade cob was easier to point, and the hay strands were easier to hide than when we pointed the regular somewhat coarser cob. It seemed to work well to recess the cob just slightly more than you want for the finished product, and then, on the same day, apply the finish cob mixture for better pointing. The fine cob, pressed into the cob base under the pressure of the pointing knife, adheres seamlessly.

We have left the panel in place for visitors to see. It is beautiful, with a lovely constellation of brightly colored bottle-ends as a design feature, as well as two large white cedar log-ends. The cob has taken on a light brown color, making the mortar in the adjacent panel look very gray indeed. The cob is quite hard, although it can be scratched — barely — with a fingernail. There are very few cracks between log-ends, and we found out that they can be removed by spraying the area of the crack with a little water and reworking the area with the pointing knife. This is best done within the first month, when the cob is still curing.

The panel is almost two years old as I write. It looks great. There is no deterioration or flaking of materials. And it has a very warm appearance.

In fairness, the panel is small — 4-feet-by-6-feet-6-inches (1.2-by-2.0-meters) — and surrounded by a heavy post and beam frame, so bearing strength of the cobwood panel is not an issue. In Tony's Roundhouse, as well as in Steen Møller's home in Denmark (described next), the cobwood walls are not called upon to be load-bearing. At this point, it seems to me that cobwood can be an attractive means of infilling a post and beam frame while making use of all natural materials — and that's pretty good. In a phone conversation in March 2002, Ianto told me that he sees no problem with cobwood as load-bearing, providing that good house-quality cob is used. Ianto, of course, has infinitely more cob experience than me.

Should there be an insulated space between inner and outer cob joints? My gut feeling is that in cold climates, the insulated space should be retained. Ianto Evans agrees. Cob builder Steve Berlant, writing in *The Natural Builder*, vol. 2, "Monolithic Adobe Known as English Cob" (*The Natural Builder*, 1999), says, "Although cob is suitable for a wide variety of climates, it may not be suitable for places with extremely cold winters. Heat loss through cob walls can be substantial." Evans points out, however, that cob will exhibit only about a third of the heat loss as the mortar it replaces in a cobwood wall. Either way, it seems to me that in addition to improved thermal performance, it is much faster to fill the middle third of the wall with insulation than with cob.

A real plus with the cobwood wall, over solid cob, is that much less cob needs to be mixed. With solid cob joints in a cobwood wall, about 40 percent as much cob needs to be mixed as with an all-cob wall. With the insulation in there, the amount of cob needed would be more like 25 percent. I am convinced that the cordwood portion of the wall has a higher R-value than does solid cob of the same wall thickness.

My thanks go to Linda and Ianto for sharing in this experience and to Ianto for reviewing this chapter. When asked if he could think of any other positive aspects of the system, Ianto commented that with cobwood it is easy to fasten things to the wall — a plus in kitchens and workshops, for example. We all agree that more work needs to be done, but we are very encouraged by the early results.

Steen Møller's Cobwood Wall, Denmark

In November 2000, at a Natural Building Colloquium in Kingston, New Mexico, Catherine Wanek presented a slideshow about her European tour of natural houses that had included a visit to Steen Møller's house in Sdr. Felding, in western Denmark. Very attractive cobwood walls were just one of the many natural building features of the home (See the Color Section.) The main construction work was done in 1996 to 1998, and the home boasts a variety of natural building techniques, including: rammed earth, straw bale, baled flax, and pressed clay brick walls; a thatched roof; passive solar features; a waterless composting toilet system; and a Finnish mass oven. In 1999, an internal wall of cobwood was built "for fun." In a phone conversation, natural builder Lars Keller, a friend of the owner, related some facts about the wall.

The non-load-bearing wall is about 10 inches (25 centimeters) thick, roughly 8-feet-6-inches (2.6 meters) high by 16 feet (5 meters) long, and features an arched doorway. The cob is solid right through the wall. The builders used a variety of hardwoods and softwoods, but the log-ends were not well seasoned and so shrank quite a bit. Lars knows that repair is as easy as wetting the wall down and re-pointing the cob, but he and Steen have not yet taken the time to do it (as of April 2002). The wall shows that the builder had a good mason's eye, both in the selection of the log-ends and in keeping consistent joints between them. Linseed oil was applied to some of the log-ends after construction, which accents the rich colors of the wood. I have seen the same technique employed with mortared cordwood masonry. The wall is a pleasing architectural feature in the home.

Lars was not actually present at the cobwood construction at Steen Møller's, but he tells me that the cob was sand, clay and "a bit of straw." The builders — all "friends of the house" — mixed the cob based on "feel." In 1999, Lars was involved with the building of four more similar cobwood walls at another home in Denmark, and he plans to continue experimenting with thicker walls that would include an insulated space between the inner and outer "cob-webs."

Chapter 21

Cordwood in Chile

Hans Hebel

(Editor's Note: This chapter has had only a few changes made since it was written for the 1999 Cordwood Conference, and these only after interviews with the author. As much as possible, I have left the "voice" of the author as he wrote it. Hans's mother tongue is German, and the household language is Spanish.)

To all the cordwood innovators at the Continental Cordwood Conference, Greetings! I am happy to report that cordwood masonry has arrived in Chile. We have had the pleasure of a visit by Rob and Rohan Roy in 1997 here at Pualafquen Valley, where I live with my large family. We have two innovations to share with you in this paper: the use of very large round and oval hollow tree trunks as giant *ventanas naturales* (natural windows); and experiments with Sika Cement Retarder.

Las Ventanas Naturales

As far as I know, we are the first to use giant hollow log-ends as windows. We first experimented with them on a practice building, during the visit of Rob and his son Rohan Roy. Later, at the addition to our house, we included some as design features in our walls, in addition to several more traditional opening windows. The *ventanas naturales* are beautiful; visitors see them and go back to their cars to get their cameras! But they shrank terribly, even after two years of drying. These windows are 12 inches (30 centimeters) thick, the same as the cordwood wall, and about 32 inches (80 centimeters) in diameter. The thickness of the wood surrounding the opening is about 6 inches (15 centimeters). As these tree trunks are made from a Chilean hardwood, it takes a long time for them to fully dry.

What to do? As we are "South Americanized," we can wait. After another year or two, we will do a repair job with new mortar and get it windproof.

The glass panes for the *ventanas naturales* were cut by a local glass installer, based upon cardboard patterns which we supplied. We used a router to make a 1-inch-wide-by-½-inch-deep (2.5-centimeter-wide-by-1-centimeter-deep) recessed border in the end-grain of the wood next to the irregular opening. Next, we simply installed the glass into the routed opening, using that simple stuff used since ages ago — putty. No silicone caulking, very expensive here. (See the Color Section.)

The new house has not passed its first winter, so I cannot say anything about the expansion of the wood in the humidity we have here in the mountains. But we have passed the first summer — the driest of the century — and have seen not only the hollow tree trunks shrink, but the larger diameter log-ends as well (those that are 12 inches [30 centimeters] and greater). The bigger the piece, the larger are the shrinkage gaps between wood and mortar. Perhaps even more drying is needed before laying the log-ends up in the wall, but it seems like they never finish shrinking. Maybe after a decade or so ....

(Editor's Note: Jaki and I and our other son Darin conducted another cordwood workshop at the Hebel homestead in 2001, and we looked carefully at the cordwood walls of the addition. Shrinkage around large log-ends and the ventanas naturales was not obvious. Hans told me that he had filled the gaps with "papier-mâché made from old newspapers mashed with a mix of half water and half waterproof white glue." This repair was almost invisible, matching closely the very light gray color of the mortar. Here is another possible solution to wood shrinkage, which could be added to Geoff Huggins' list in Chapter 12.)

We did use a small amount of framing wood that was not 100 percent sound; that is to say, it had some pre-existing deterioration or wood rot. We will not make this mistake again. Two of the ordinary window frames (not the *ventanas naturales*) have had a little termite attack. We will have to replace the pieces. It is better to check carefully before installing any wooden pieces, including window frames, posts, or log-ends.

Another error we committed was choosing the pillars (posts) not strong enough. We used pillars made of 2-by-10-inch planks, but we should have used 4-by-10-inch material. The larger pieces would be less likely to deteriorate and will still have enough strength after decades. Ours, perhaps not.

21.1: An historic moment, February 1997. Carolina Salvado places the first log-end in Chile—perhaps in all of South America—while her mother and sisters look on. Credit: Rob Roy.

Sika Plastiment™ Cement Retarder

We have some good news to report: our experience with the liquid cement retarder that we used in Rob Roy's regular mortar mix (9 parts sand, 3 parts soaked sawdust, 2 parts Portland cement, 3 parts lime). Even though we had very little experience in cordwood masonry — only the work we did with the Roys on the practice building two years ago — we had just two or three small mortar cracks in the entire house. The stuff seems very good. It is called Sika Plastiment™, a "water-reducing, retarding densifier." The company gave us a sample free for the asking, just to try it out. It is a yellowish, milky liquid, and the normal mix is between 0.6 percent and 1 percent of weight of the cement. But we put in about 1.6 percent to compensate also for the lime.

(Editors Note: To find out where to get Plastiment™ in the United States, send an email to: eaton.paul@sika-corp.com; or call or contact Mike Campion, Marketing Department, Sika Corporation, 201 Polito Ave., Lyndhurst, NJ 07071. Phone: (800) 933•SIKA or, from outside the US, (201) 933•8800.)

Even with all the errors committed, we are happy with our new little home, which is connected to a framed house that we had built a few years earlier. As far as we know, this is the first cordwood house in Chile. Now, I have heard of another cordwood home being built about three hours from here. I have not visited the place, but think it is three stories: the first in cordwood, the second and third in wood frame construction. A lot of others will follow in the upcoming years, as we are working on a complete little ecological village that will give work to Mapuche (native South American) families living in the area.

In 2002, we are completing a new cordwood house for one of the members of the Eco-Village here in the Pualafquen Valley near Conaripe. We have returned to the mortar mix that Rob and his family have taught us at two workshops here, without cement retarder. The cement retarder was a successful experiment, but it is not available in the common market, only in barrels of 60 gallons (230 liters). Also, we prefer to avoid the use of chemicals for many reasons.

There will be many new cordwood homes in the Eco-Village. We are now collecting many of the giant "natural windows," starting a whole new industry for

21.2: Hans Hebel (left) and Rob Roy help pass one of the giant ventanas naturales to waiting helpers.
Credit: Rohan Roy.

families to do in the wintertime. Except for these wonderful windows, I would not take logs bigger than four inches (ten centimeters) in diameter. There is too much shrinkage with bigger logs of our Chilean hardwoods.

Cordwood masonry is perfect for people who want to work with Mother Nature. It looks great here in the southern woods of Chile, and we hope to inspire many folks around here. I do not have much to add to the facts already given, but I hope you will like the photos attached. And if you come to Chile, drop us an email at: inmuebles@telsur.cl before your arrival, and we will take a little care of you. And you may also look at our new website at: www.pualafquen.com.

CHAPTER 22

One Old and One New in Sweden

Olle Lind

One Old

In Botkyrka, near Sweden's capital of Stockholm, there is a small cordwood house, deserted since 1950. The house measures 25-feet-by-16-feet-5-inches (7-by-5-meters) and has a wall thickness of about 12 inches (30 centimeters).

It has been documented that the building was constructed about 1860. I am somewhat surprised that it is still standing, since the foundation was badly constructed from the very beginning and consists only of stones placed directly on the soil. On top of the stones are beams (or sill plates), the same thickness as the cordwood walls. Above the sill plates, the walls are constructed of split spruce log-ends, with the masonry courses stabilized by horizontal wooden planks. As Swedish spruces have low resistance to rot, I found it remarkable that most of the log-ends are unbarked and are still sound.

The mortar between the logs is clay without Bentonite (naturally occurring expansive clay), reinforced with about 40 percent sawdust. There appears to be very little mortar shrinkage.

And One New

When I retired a few years ago, I decided to find a better place for my bird-breeding activity, a place where I could settle down with my cockatoos and give them space without disturbing grumbling neighbors. I left Sweden's south coast and moved far north, into the woodlands. It wasn't that easy to find the right place where I would belong and prosper. In less than a year I moved four times, got married and divorced, and all the time I carried more than a hundred cockatoos as hand-baggage — an ordeal, perhaps, worthy of the *Guinness Book of World Records*. But this was a bit tiring and I lost some money. There's no profit in breeding almost extinct cockatoos, since trade with this type of parrot is prohibited. And if I am lucky

22.1: This old house near Stockholm, Sweden dates back to about 1860. Despite a foundation of stones on the ground, unbarked spruce log-ends, a clay and sawdust mortar, and more than 50 years of neglect, the walls are still intact. Credit: Olle Lind.

22.2: Although more than 140 years old, the unbarked spruce log-ends of this house in Sweden are still in good condition. Note the use of horizontal boards as stabilizers between courses. The mortar is a mixture of clay and sawdust. Credit: Olle Lind.

and the birds get babies, I need more space. At the same time, I got to know about Rob Roy and modern cordwood masonry housebuilding and decided to give this building method a try.

The local building inspector, of course, didn't know anything about cordwood buildings, even though there are said to be hundreds of old examples in Sweden. But finally, he studied Rob's *Complete Book of Cordwood Masonry Housebuilding: The Earthwood Method* (Sterling, 1994) and said, Okay.

I decided to build in three sections and to conclude one at a time, since I really needed the space. If I wasn't a bit crazy, of course, I would have chosen some kind of conventional modular building units. For the first section, which measures 19-feet-8-inches-by-32-feet-10-inches (6-meters-by-10-meters), I chose Swedish spruce, which is locally available. It dries out easily but also absorbs moisture fast. The shrinkage percentage for Swedish spruce is: radial, 4 percent; tangential, 8 percent; volumetric, 12 percent.

Barking the logs took a very long time. Working with the birds took me more than 12 hours a day, leaving just a few hours for the barking work. The spring weather was unusually warm and dry, and in less than a week the bark was extremely reluctant to let go of the log. As a result of my experience, I ordered barked wood from a pulp mill for the remaining sections of the building.

To begin with, I built the post and beam framework, including the roof. I used box posts of 8-by-8-inches at the corners, with posts of 2-by-8-inches along the side walls, spaced every 6-feet-7-inches (2 meters). All posts were placed in line with the outside of the walls. Due to the dry weather, the logs had dried out too much.

The next year, when I began to lay up the cordwood walls, the weather changed completely. It started to rain and continued day after day, all through the summer. The mortar didn't dry. The logs absorbed moisture from the high humidity and from the wet mortar. The mortar cracked from log-end to log-end. This was not to my liking, of course, so I tried to reduce the amount of water to minimize the shrinkage.

I started with a strong mix (8 parts sand, 4 parts sawdust, 2 parts Portland cement, 3 parts lime) and finished with a mix of 15 parts sand, 8 parts soaked sawdust, 2 parts Portland cement, 5 parts lime, and some dish-washing liquid — about a half-coffee-cup per wheelbarrow load. The cracking of the mortar turned out to be about the same, whatever the mix. The walls expanded approximately ¼-inch per 3 feet (5 millimeters per meter) with the first mix and about ½-inch per 3 feet (15 millimeters per meter) with the last mix. The result of this expansion was that, unintentionally, the walls became load-bearing and even managed to tear off a couple of bolts in the post and beam framework.

Around split logs, I noticed cracks in one to two weeks. Between round logs, cracks would appear in one to two months, which was similar to the time the mortar needed to dry. I assume that at least some of the cracks were formed as early as during the first days, but didn't appear due to the structure of the surface. The slow drying of the mortar and the fact that the log-ends continued to expand all summer was probably caused by the extreme weather conditions, consisting of almost constant rain and/or an atmospheric humidity of between 80 and 100 percent. The ongoing wood expansion didn't stop until I began to heat the inside of the building.

The experiments with washing liquid in the mortar (in order to reduce the amount of water needed) were dramatic, with both good and bad results. In the first mix, I reduced the amount of water about 40 percent, but in the last mix by only about 15 percent. It was easy to stir the mortar to a creamlike state, but it solidified rapidly, which led to difficulties in getting the next handful of mud to stick to the previous one. And the expanding log-ends easily caused cracks when they were put in the mud.

I tried increased amounts of filler (sawdust and sand) in an attempt to reduce mortar shrinkage, but still experienced similar problems.

It was a challenge working alone ten feet (three meters) above the ground: laying up a little mud, climbing down, walking around the wall, climbing up the other side, laying up a little mud, climbing down, etc. I insulated with sawdust and lime, at about a 12 to 1 ratio. All this took me a very long time, so I had to

22.3: The south side of Olle's cordwood building. Credit: Olle Lind.

use Cemtex retarder in the mortar. If I used too much retarder, I was still able to do the pointing the next day. The Cemtex company advises a normal rate of use between 0.3 percent and 1.5 percent of the cement weight, with a maximum of 3.0 percent. An increase of 0.1 percent retards setting by an additional hour at 65 degrees Fahrenheit (18 degrees Celsius).

In spite of all the trouble with the weather, the cracks, and the wood expansion, I am very pleased with the result. I enjoy the look of the wall's surface, and most cracks are only of the hairline variety. But for the next section, I have, after some experiments, decided to try a water-repellent agent, *zinkstearat* (zinc stearate), as an admixture.

The preceding was written for the 1999 Cordwood Conference in New York. Since then, after several summers' work and a lot of help from my new wife, the house is finished. And we are very pleased. I am convinced that it is impossible to fail with a post and beam framework with cordwood masonry as infilling.

Rob Roy's Comments

It is difficult and not entirely reliable to diagnose something from 4,000 miles (6,400 kilometers) away. But a clue to Olle's mortar cracking may be found in his observation that cracks appeared around split log-ends in one to two weeks, while cracks took one to two months to appear around the rounds. This would seem to indicate a case of wood expansion from water absorption, rather than mortar shrinkage: the result is still cracks in the mortar web.

Moisture absorption into wood varies greatly, depending on the direction it comes from with respect to the wood grain. Wood dries up to ten times faster through end-grain than through side-grain, which is why even debarked logs left in long lengths — say, 8 to 10 feet (2.4 to 3 meters) — do not dry very quickly. Transpiration of moisture is a two-way street. If wood gives off its sap moisture rapidly through end-grain, we can be sure that it will take on moisture through end-grain almost as rapidly. The exposed longitudinal fibers of the wood act almost like straws or capillary wicks. In our cordwood saunas, we pour water over the hot stones that surround the stove, converting it instantly into *löyly* (LOW-loo), the sacred steam. Within a couple of minutes, the dry log-ends (all on end-grain) absorb this high humidity, and the atmosphere becomes dry again, awaiting another infusion of *löyly*. Fortunately, when we build a normal cordwood wall, the end-grain never comes into contact with the moisture in the mortar. If it did, we would have even greater problems with wood expansion.

The next most hazardous threat of rapid moisture absorption is if we place wet mortar up against sawn logs, as my son Rohan found out when he helped build a cordwood house in Memphis, Tennessee. The longitudinal cells are ripped open during the sawing and act very much like end-grain: capillary action ensues. Almost all of the "log-ends" at the

Memphis house had sawn sides — pieces of scrap timbers mostly — and Rohan and his employer laid them up like concrete blocks, with mortar that was much too wet for Rohan's liking. The result was similar to that experienced by Olle Lind: the expanding wood actually lifted up on the upper plate beam, which was well fastened to the top of the posts. Some of the posts were lifted right off the foundation, until the builders realized that they had best not fill in the final, topmost mortar joint beneath the plate beam until the wood had finished its expansion.

At a cordwood workshop in North Carolina in 2001, Jaki and I helped our hosts build a very attractive pump house using stackwall corner quoins. Charles Shedd, our host, made the quoins by ripping the top and bottom slabs off of tulip poplar logs, leaving 6-inch-thick (15-centimeter-thick) timber "milled two sides." These timbers were then cut into 12- and 16-inch (30- and 40-centimeter) quoins and set to dry. We prepared all the quoins the day before the workshop by applying a single coat of Thompson's Waterseal™ to the sawn edges of each quoin. We also set a few roofing nails into the edges of the quoins that would be in contact with the mortar, to greatly improve the friction bond between quoins and mortar (see Image 22.4).

22.4: Stackwall corner detail at Charles and Andrea Shedd's pump house in Bakersfield, North Carolina. Roofing nails were installed an inch or two (two to five centimeters) from the edge of the quoin to improve the friction bond with the mortar. Credit: Rob Roy

Although the tulip poplar log-ends shrank quite a bit — Charles is now using Geoff Huggins's method of stuffing gaps (see Chapter 12) — there were no problems with expansion or contraction with the quoin material. The waterseal, we feel, is a cheap insurance against potential wood expansion — certainly with sawn log-ends and perhaps with split log-ends, too. Apply and dry such chemical-laden materials out in the open. In general, I am against the use of preservatives, sealers, chemicals, coatings, or stains on log-ends. They are almost always unnecessary and environmentally suspect. The situation described here, though, is my exception to the rule.

When wood is split instead of sawn, there is much less exposure of the longitudinal fibers or cells. However, the split edge does give moisture an access route between adjacent cells, increasing hydrostatic pressure and wood expansion. Consider that split firewood dries very much faster than non-split wood. Drying and taking on moisture are simply two sides of the same coin.

With the round log-ends, cracks in Olle's mortar took about four times as long to appear as when he built with split wood. I submit that this is because water absorption is very slow through the outer or epidermal layers of the wood. If these outer layers were not waterproof, trees would "leak" — which is, in fact, what happens when we violate their "waterproofing membrane" with an ax or, say, tap maple trees to derive the sweet sap.

To summarize:, the transference of moisture in and out of wood is greatest along end-grain, next greatest along sawn surfaces, less but still significant along through split surfaces, and least through the wood's outer epidermal layers (which are seen on the end of a log as annual growth rings).

22.5: Transpiration of moisture into wood. To consider drying of the wood, just think of the arrows pointing in the opposite direction. Credit: Rob Roy

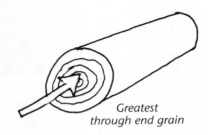

Greatest through end grain

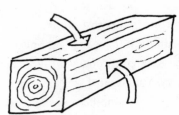

Next greatest through sawn side grain

Less through split side grain

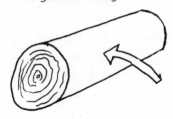

Least through epidermal layers

CHAPTER 23

Creating with Stone, Wood, and Light

Tom Huber

In the spring of 1996, we purchased a 20-acre (8-hectare) parcel of land in southwestern Michigan where we could begin work on developing our homestead. The property features rolling hills, with wooded ridges of pine and assorted species of deciduous trees. We also found several good outcroppings of fieldstone. With a good crop of stone and pine we had the beginnings of our building materials for stone and cordwood structures.

Prior to our construction projects, we put in an orchard of 38 fruit trees and established a tree and shrub nursery. Every year we replant some of the nursery items to create more forested areas for wildlife habitats, scenic beauty, and cordwood production.

Before the end of the first spring, we also felled over a dozen red pine trees, purchased a load of barn beams, and hauled several loads of utility pole pieces and fieldstones. All of these materials were eventually recycled into our first couple of building projects.

After reading most of the available literature on cordwood and stone masonry (slip-form style) and conversing with a host of builders (cordwood and conventional), we were ready to begin building. Especially helpful in this early preparation phase were the writings of Rob Roy, Richard Flatau, and Jack Henstridge on cordwood masonry, and Helen and Scott Nearing on slip-form stone construction.

By the end of the first year we had erected two stone and cordwood hybrid structures. The first structure was a small tool shed, using a post and beam frame with 10-inch (25-centimeter) stone and cordwood walls. The second structure (which we call the lodge) was also built post and beam style, with stone and cordwood infilling. We added a solar porch to the lodge the second year, and a tractor lean-to was attached to the small shed. We also harvested over 100 pine trees for future cordwood projects. They were skinned and cut up into 16-inch (40-centimeter) log-ends and stacked to dry by summer's end.

In 1998 we added a passive solar house to the west side of the lodge. After designing many different homes, we finally built one to best use the southern exposure of our site. All of our buildings have white metal roofs to reflect the sun's heat, collect rainwater, and shed snow. We made use of three essential building ingredients (stone, wood, and light) in our construction projects.

The Art of Stone Masonry

I'm not sure when it started to happen, but for as long as I can remember I have been enamored of stone masonry. I find good stone work to be visually pleasing, an esthetic work of art. My eye is especially attracted to fieldstone masonry having lighter mortar joints. The white joints tend to bring out more of the color from the stones. A solid stone foundation for a barn, house, or simple outbuilding suggests strength and permanence. Even a little stonework in a structure can add color, texture, and character.

Some years ago, I was introduced to the work of Helen and Scott Nearing. I read with interest about their accomplishments in homesteading and simple living. Most of all, my attention was captured by their stonebuilding projects. I read all of their books, paying special attention to their use of the slip-form method of building with stone. Later, I visited their Forest Farm in Maine and saw firsthand some of their stone projects. I was hooked. I knew that in some way I would have to feed my stone addiction.

Being a complete novice, I was hesitant to build an entire house of stone. I was constantly attempting to figure out an easier way to construct a beautiful structure, without having to endure all the labor of lifting and placing so many stones (and forms) so high off the ground. It was about this time that I came across a book on cordwood masonry. The photo on the front page showed a structure that looked like it was made of stone. Something clicked inside me. I'd found the missing piece!

After more research and contemplation, a creative synthesis was born. I would combine stone with cordwood masonry to create beautiful, solid works of art. I would use my own adapted forms to create a 2-foot-high (0.6-meter-high) stone wall, and then use cordwood masonry to finish my walls the rest of the way. This would be done within a post and beam framework, where I could attach my plywood forms directly to the posts. Using this method I would only have to build a few forms 2 feet high to build the low stone portion of the wall. By finishing the wall with cordwood, I would be eliminating the bulk of the labor required if I were to continue the wall with stone. I figured it would be much easier to position log-ends one at a time than to continue raising the slip-forms along with all the stone and concrete to fill them. The two-foot-high stone portion of the wall would provide stability and raise the cordwood portion higher off grade, where the wood would be less prone to rot.

23.1: Tom and Holly Huber's test building, a tool shed, built in Dowagiac, Michigan. As it is less than 100 square feet (10 square meters), no building permit was required. Credit: Tom Huber.

Once I knew that I would be building with stone, I attempted to apprentice myself out to a *bona fide* mason, but it never panned out. My only alternative was to start mixing mud and augment my book learning with actual experience. This approach worked for me and I would suggest it to the reader. I highly endorse the slip-form method, which enables even the novice to build strong and beautiful foundation walls. Try to read as much as possible on the subject first. Learn from the trials and errors of others before making your own mistakes.

Stone masonry takes time to do a good job — even more time than cordwood masonry — but there is nothing more satisfying than to gaze upon a creation made of stone.

Rubble Trenches and Frost-protected Shallow Foundations

Frank Lloyd Wright was fond of using trenches (below the frost line) filled with stone rubble instead of solid concrete for foundations. Any water would easily drain through the stone and out drain tiles rather than potentially collect, freeze and cause damage through heaving in colder climates. On top of the rubble trench a reinforced concrete grade beam was poured to serve as a footer from which the rest of the house would be built.

We used a rubble trench foundation on our sloping site to create an even top surface for our grade beam when we attached our house to our previously built lodge. This allowed us to tie-on to the existing block foundation wall without having to build a stepped concrete foundation for the house. The site sloped down approximately 50 inches (130 centimeters) from the outside wall of the lodge to where the house would end. If we honored the frost line requirement (42 inches [107 centimeters]), it would have meant putting in a foundation over 7 feet (2 meters) below grade. We did not want to put in a basement, since we had already decided on going with an insulated slab for maximum solar gain (see below). Even if we had decided to build a walk-out basement (which many suggested), it would have had to have be a small one, so as not to disturb the footing of the existing lodge. We would have had to keep some distance to prevent disturbing the mostly sandy soil upon which the lodge was founded. For all these reasons the rubble trench concept turned out to be the best solution, especially price-wise. The book, *Foundations and Concrete Work* (The Taunton Press, 1998), has an excellent article on the rubble trench method.

Nothing complements the beauty of stone better than the beauty of wood. In our first few building projects, we had used some rough-sawn pine siding from a mill, and I knew right then that I wanted to incorporate more rough-cut materials in our timber framed house. We ended up ordering all the posts, girders, loft floor joists, stairway stringers and stairs, wavy-edged siding, and trim boards from a sawmill. Most of the materials were kiln dried for stability. It was still the best bargain price-wise and beauty-wise of the whole project. It perfectly complements the overall rustic theme of building with stone and cordwood. The Douglas fir flooring in the loft also communicates a simple, natural beauty of an earlier time.

Cordwood Confessions and Suggestions

After poring over the available cordwood literature for over a year, it was finally time to roll up my sleeves and get to work. Even though I thought I knew the essential aspects of cordwood masonry and obstacles to avoid, I still ended up making my fair share of mistakes. Listed below are the lessons I learned.

1. Understand the species of cordwood that you choose to build with. Let it dry properly and form checks on the end-grain before you put it in your walls. Split larger pieces to reduce shrinkage. Consider building with cedar, if at all possible. It is the best choice for its stability, R-value, rot resistance, and beautiful aroma; and it is a sheer pleasure to work with. If I were to do it all over again, I would make even more treks to northern Michigan to harvest northern white cedar for all my cordwood projects. I have a good cache of it now and will scrounge any downed cedar trees I see in our area. I also had good luck getting some leftover cedar log-ends from a log home company, and other logs from a tree removal company. If cedar is impossible to locate, at least use a softwood species of wood.

2. Recycled utility poles work well for cordwood projects, especially if they are made from cedar. However, we're not all as fortunate as Cliff Shockey, whose recycled utility poles were untreated above grade (see Chapter 4). Most utility poles are chemically treated and should only be used in uninhabited structures such as outbuildings. Utility poles are generally free from utility companies, are for the most part sound, and require no debarking. Even though they may seem to be quite dry, it is still good to wait at least a few weeks after cutting them to size before you put them in a wall. If you don't, you will most likely have some log-end shrinkage. Even log-ends cut from timbers decades old will shrink some. When this happened to me while building our small tool shed, I shared my puzzlement with Rob Roy, who told me I should have waited at least 30 days before placing them in a wall. Of course, it makes perfect sense that even old logs will release some moisture when cut up into smaller pieces.

3. If you have any doubts about whether a recycled barn beam or antique log-end is of sound quality, reject it without a second thought. Take your time and build your

structures as though they will last for over 200 years. Shortcuts rarely pay. The two-foot stone wall strategy and ample overhangs on your buildings will really make a difference over time on preserving your cordwood walls. Planned lean-tos, added porches, and other add-on strategies that help keep cordwood walls protected from the elements will insure that your works of art will outlast you.

4. We found drawknives to work best when skinning pine logs. Springtime is usually the best time of the year to skin logs, when the sap is rising. Do not delay skinning after cutting, especially later in the summer when bark tends to get baked on the logs. Cedar bark tends to pop off easily with a thin chisel and can then be pulled off in strips.

5. Buzz saws can really help make short work out of bucking up logs into log-ends. These large (24- to 30-inch-diameter [60- to 76-centimeter-diameter) circular saws are often made to connect to the "power take-off" (PTO) of a tractor. We purchased a PTO-driven buzz saw from a farmer for only $25. A lot of these antique saws can be found in the back of old barns and storage lean-tos.

6. Spend sufficient time tuck-pointing your walls. Tightly push mortar against the log-ends so that it forms a 90-degree angle or perpendicular plane to the log-ends. Resist the urge to just smooth out the mortar. Any thin sections of mortar sloped against log-ends dry out rapidly and become flaky. Removing these later takes time.

7. Experiment with different mortar mixes after you read about the various recipes in the cordwood literature. I've had good luck using the same simple mix that I use for my stone pointing (three parts sand to one part white masonry cement). I have also been experimenting with PEM or Paper-Enhanced Mortar and have had exceptional results with this new mortar medium. (For convenience of comparison, my personal PEM recipe is given in Chapter 15.) Once you decide on the best mix, use it continuously for both uniform appearance and performance (color, texture, strength, etc.) Stick with the same manufacturer, to avoid color variations.

8. If using lime-treated sawdust for insulation, prepack it between log-ends to prevent voids from settling later on.

9. I used Perma Chink™ on the mortar walls of our lodge, with excellent results. Although it was time consuming, I had the best results applying it with a small soldering brush. Next, the thin layer of Perma Chink™ (or Log Jam™) should be smoothed with a pointing knife. Not only did it brighten up my walls, but it also tightened the mortar sections, making it more weatherproof (see also Chapter 12).

10. Post and beam frame. I generally prefer to erect the frame of the building first and then do stone and cordwood infilling. This way, you can get a roof over your head

before you lay up any of the log-ends. By keeping the direct rain and sun off your work, you can put up your walls in any weather conditions. This approach might also serve as a deterrent to excessive log-end shrinkage by providing shade from the hot overhead summer sun.

11. Hired help. The post and beam approach may also help to organize the work stages of your project and provide a division of labor. For example, you may need to hire some help in getting the overall frame erected and the roof put on, and then you can more easily go about building the walls as time permits. This approach is particularly valuable if a building project will take more than one building season.

By no means do I consider myself to be an expert on cordwood masonry. I've had enough experience with the medium to know how much I like it and to know what mistakes to avoid next time. I have a list of future cordwood projects I hope to complete someday, but I don't share this with my wife. Cordwood masonry takes significant time to do the job right. I can promise you, though, that if you allow for the time it deserves, you'll make a lot fewer mistakes and enjoy the process and finished product even more.

Let There Be Light

I'm not alone in encouraging the prospective builder to get to know their building site as well as possible. While driving, I am almost constantly scanning the roadside and hillsides examining the man-made structures and homes in the area where I live. I'm often surprised by how frequently a building site's most prominent features are not fully exploited. In particular, I rarely see where a site's southern exposure is taken advantage of for solar gain. This serious oversight can add significantly to a home's energy cost and make for a darker interior. Most construction projects, at least in the north, can benefit from some southern solar exposure.

When we initially started planning our building projects, I thought for sure I knew exactly where the house was going to be built. It seemed so obvious. After two years of completing pre-house projects and getting to know more intimately the land we had purchased, I recognized that my first decision on house placement was wrong. I had not carefully considered the prevailing winds and the excellent southern exposure (with conifer trees planted to the north) that our land provided. It was for this reason that we ended up building a passive solar house attached to the stone and cordwood lodge.

We used the classic saltbox design (tall south wall) to incorporate a lot of glazing to the south for passive solar gain. The overall open floor plan was kept quite simple by using a post and beam framework. We also made sure there was plenty of thermal mass, in order to keep a steady (as well as comfortable) temperature. Not only are some of the beams massive, but there is also plenty of mass in our insulated concrete slab, the stone and cordwood masonry, and a fieldstone-faced masonry stove in the center of the house.

Solar Heating and Cooling

The proper amount of mass provides crucial storage capacity for solar gain. With too little mass, the house simply overheats on sunny days, even in the middle of winter. As cordwood builders will testify, the mass also helps absorb the heat of the summer, keeping the interior air at a lower temperature. At night, proper ventilation and cooler temperatures help discharge any extra stored heat so that the whole process can begin again the next day.

We installed a power vent in our walk-in attic and an exhaust fan in an overhead attic, which help to ventilate the house. We also made extensive use of radiant barriers (foil-faced foam and bubble wrap) to keep out the sun's radiant heat. A properly calculated overhang can also help keep out the direct summer sun without impairing maximum solar gain in the colder months (see image 4.1).

For maximum solar gain, a concrete slab can be stained a darker color, or a pigment can be added in the concrete prior to the pour. We had moved into our new house before we finished the floor and used a special stain that interacts with minerals in the concrete. Every application therefore creates a different, varied appearance, and it cannot chip or peel off. (See www.kemiko.com for product information.)

23.2: The masonry stove at the Huber house.
Credit: Tom Huber.

Masonry Stove

A masonry stove provides perhaps the most efficient and comfortable form of wood heat known to man, while at the same time serving as a beautiful focal work of art. It is also the perfect companion to solar heat. Its large thermal mass provides a great capacitor for solar heat storage. Rob Roy claims that his 23-ton (27-metric ton) masonry stove actually works in the summer, as well, by "storing coolth."

A well-constructed passive solar house is easy to keep heated in the spring and autumn months by the power of the sun alone. The masonry stove may be used only occasionally. In the winter we use the masonry stove as our primary heating system, but on good solar days we may not need to fire it up at all. We installed an in-floor hot water radiant system for backup heat if we needed to be away for an extended time in the winter.

Thermal-shutters

Many solar books mention the importance of proper insulation (including some form of insulation for large areas of glass) to hold in the heat at night. Residents of solar homes more often install insulating shades or shutters over their windows than turn up the thermostat when cooler weather is approaching. We built simple insulated shutters by cutting foil-faced sheets of polystyrene foam to fit our window openings. We then lined the foam-exposed edges with a durable radiant tape and covered the sides with contact paper of a pleasing design. Our shutters are simple but effective.

Closing Thoughts

Writing these words has allowed me to think back on all our building projects and the decisions that were made along the way. Anyone who engages in building projects will inevitably wish that they had done something different. I am no exception to this rule. So I offer the following wish list to prospective owner-builders.

I wish I had practiced more fully the sage advice of Scott Nearing, who once offered the wisdom of, "Pay as you go." We were fortunate not to go into debt until we were into our house project, but no debt is still better than some debt. I can rationalize the decision to take on a house mortgage by quoting the low borrowing rate, but the truth is that if we had been more patient and had hired less skilled labor, we may have been able to build mortgage-free. I, however, was a complete building neophyte when we began, so my skills were limited. We also have two small children, which would have made living in a shed for an extended period of time beyond stressful. Just ask my wife. She tolerated enough of my absence from home as it was.

Chapter 26 has some good suggestions on avoiding a mortgage, some of which we used to keep our mortgage down. In particular, consider going at the project in stages, as Rob suggests with the add-on strategy (my personal favorite). Design a structure that will allow for expansion through additions without becoming a hodgepodge of a house when finished. The first relatively completed part of the structure should be habitable while the next part is being worked on.

I think it is crucially important to make an honest assessment of your skills, assets, and resources and do the best you can. Without setting limits, I know my marriage would ultimately have been even more taxed than it already was. Thus, we decided to finish the majority of the house in one building season, which simply required more money. We could have drained all our assets to eliminate a mortgage, but as my wife instructed me, we were getting a higher return on those investments. Fortunately, we practice a moderate version of what Charles Long (*How to Survive Without a Salary*, Warwick, 1996) calls the "Conserver Lifestyle" in our attempts to reduce spending, so we hope to pay off our mortgage as soon as possible. Everyone's situation is different, and it is important to take a realistic inventory prior to buying the land, materials, tools, etc.

With the preceding in mind, I also recommend that one start preparing to build as soon as the decision to build is made — hopefully, before one has children and a lot of worldly responsibilities. We started by building a small tool shed to keep our tools protected from the elements. (We used a small trailer for them initially while we built the shed.) Next, we built what we now call the lodge, to store all our building materials for the house and to provide a place to eat and sleep intermittently while building the house. This is how the pioneers generally approached the tasks of homesteading. We were fortunate to be paying cheap rent close by on a lake-house property that we helped manage. If we had had higher rent to pay, we may have been more motivated to live more completely on our land while we built.

I've been told that you have to build three houses before you finally build one right. I'm not sure if I will ever build three homes, but I do plan to continue with various building projects into the foreseeable future. Nothing in my life thus far has been as satisfying as creating with stone, wood, and light.

CHAPTER 24

The Community Round House at Pompanuck

John Carlson and Scott Carrino

IN 1986, AS A GIFT MEANT TO OPEN THE EYES of a conventional stick frame builder, we came upon one of Rob Roy's early books on alternative building, which described the construction of his two-story round Earthwood house. Little did the giftgiver realize that cordwood construction would become an obsession. After a visit to Rob and Jaki's Mushwood Cottage in 1987, and then a visit by all of Pompanuck Farm's community members to Earthwood in 1990, we decided that this building technique was appropriate for our purposes. In October 1990, we began the process of imagining and designing our community center in Cambridge, New York (see the Color Section).

Two of our group had been professional designers and builders, and others had various renovation and interior design experience. Many good artistic eyes and talented hands gave us the confidence and enthusiasm to attempt a large project. The responsibility of overseeing the design and building the structure went to the two experienced builders. We were all committed to creating a working environment where everyone involved was valued, their ideas appreciated, and their new skills nurtured. We felt strongly about having a non-gender-biased work environment, and for the most part we succeeded. There were periods when, because of time and seasonal pressure, we fell into the traditional roles of women doing the cooking and providing childcare and men doing the building. During these times we met as often as possible to air our concerns and tried to create solutions to any frustrations that arose.

Designing the Community Center

The interior design process involved the six adults and two children of the community in various ways. Getting everyone to agree on a floor plan and structural design was, on the surface, a lot easier than we thought it would be. On a deeper level the task was much larger than anyone imagined. We simply did not know what problems to expect or how large an undertaking we had embarked on. The task of layout was challenging, because the building had to be designed to be as flexible as possible to accommodate other uses down the road. And it had to become the full-time home for two families until the community increased in members, when separate homes would be needed.

There were many sessions where we gathered around a table with a three-foot-square (one-meter-square) piece of cardboard, on which we scaled a 48-foot-diameter (14.6-meter-diameter) circle drawn to represent the floor plan. With two stories, this meant 3,620 square feet (336 square meters) of gross area, almost exactly 50 percent larger than the Earthwood model that had inspired us.

Scaled house components were cut out of paper, representing everything from tables and beds to kitchen and bathroom fixtures. As these elements were placed, they began to define the rooms. This cardboard model took on the quality of a giant board game. In fact, on New Year's Eve, we were moving champagne glasses around to the various rooms on the floor plan, dreaming with each move about the activities that would take place there.

We created a list of parameters that we developed by visiting other cordwood and cordwood-inspired houses in the Northeast. From those experiences we assembled details that would work well with our community situation, our county building codes, and the contours of our land. We knew we wanted to shelter a good portion of the bottom floor with an earth berm, and we wanted an earth roof to minimize environmental impact. The building would

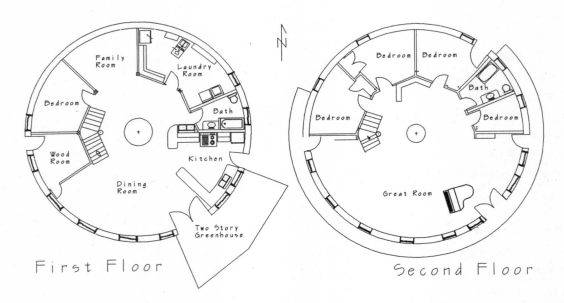

24.1: First and second floor plans of Pompanuck Farm Community Center in Cambridge, New York.

be heated mainly with a central, wood-fired Finnish style stove. Large windows would face south, with appropriate overhangs to take advantage of passive solar gain.

As the design process continued, other important elements became clear. Because of the large size of the downstairs space — with more than half of it being open plan and below grade — we chose a 9-foot (2.7-meter) ceiling height to keep it from feeling dark and cavelike. We incorporated recycled materials, such as 100 single-glazed, double-hung windows obtained for the cost of shipping them from the renovation of a New York City apartment building. These windows came in a few sizes, and we chose the ones that went well with the proportions of the structure.

Structural Considerations

Code required that we hire an architect or engineer to review the structural elements and stamp the plans for our large building (see Chapter 29 for code considerations). We were surprised when our architect suggested that we use a steel frame to halve the span between the outer walls and the load-bearing cylindrical masonry mass at the center. At first, we met this recommendation with a fair amount of narrow-mindedness. We had a long-standing relationship with wood and were going to use only "natural" materials in this building.

It did not take much convincing that steel was a sensible option, after the architect ran calculations on the wooden joists and rafters that we thought might work. We had not considered the potential effects of a 100-inch (254-centimeter) snow load on a relatively flat, water-saturated earthen roof. The supporting girders would have to be far more substantial than we had anticipated. Steel was an even more obvious choice where, to create an 18-foot (5.5-meter) open span, we wanted to remove one post of what would have been a perfect octagonal framework. We had thought that a white pine girder could carry this load, but the architect's analysis showed that a 14-by-14-inch clear white oak girder was needed. Such a girder would weigh over 1,700 pounds (771 kilograms), whereas the required steel I-beam would weigh less than 800 pounds (400 kilograms). By choosing steel, we would save our backs as well as our pocketbooks.

Once we had excavated the site, we began to remove material for the foundation. All of the soil conditions to this point were beautiful sandy gravel — the stuff a builder loves to find under a house site. Our rejoicing was short lived. Upon digging on the north side of the site, we found a vein of pure clay (a potter's dream). When we notified our architect, who had already told us that we would need a substantial foundation to support the weight of our structure, he increased the size of the footings from 36 inches wide by 10 inches deep (91 centimeters wide by 25 centimeters deep) to 48 inches wide by 16 inches deep (122 centimeters wide by 40 centimeters deep). Over 27 yards of concrete were consumed by these monster footings.

Finding and Harvesting Cordwood

Our search for cordwood led us to our own backyard. Our land borders 1,200 acres (485 hectares) of multi-use state forest. We contacted Ron Cadieux, the forester in charge, to find out what it would take to harvest trees there. We had the choice of red pine or Japanese larch. The pine was lighter and probably would have given us a slightly higher R-value. The larch — a moderately heavy and hard deciduous conifer — has a similar specific gravity to that of yellow pine but is more rot resistance than red pine. We decided to go for the resistant quality of the larch over the insulation value of the pine.

We contracted to harvest 20 full cords of wood which worked out to be about 68 trees that were 8 to 13 inches (20 to 33 centimeters) in diameter and 80 to 100 feet (24 to 30 meters) tall. The groves had been planted 35 years previously and contained trees that were straight, tall, and relatively knot-free. We would be responsible to execute a "select cut" harvest of the trees. Ron was able to select trees that were not the straightest or the best for lumber, but would still fill our needs for cordwood. We also used trees of a smaller diameter, including some that were of no use for making conventional lumber.

Ron recommended tree farmer and neighbor Bill Bagley to help us harvest the trees, confident that Bill could take down large trees in a plantation without injuring himself or the remaining trees. Timing was critical. We needed to wait until mid-May, when the sap would still be running, making it easier to debark the trees. However, at this time, trees not being harvested risked injury as falling trees scraped past their sensitive new springtime bark, leaving them vulnerable to insects and decay. For this reason, harvest is not usually permitted until after the trees go to bud. Ron made an exception in our case, provided we did not damage standing trees. Our care in this harvest cannot be overstated.

One fine May morning, the silence of the forest was broken by the buzz of chainsaws, and the harvest began. Bill wielded a chainsaw with the grace of a surgeon, dropping a series of trees so that we could begin removing the bark. We used truck leaf-springs found at a junkyard as our debarking tools. Their natural curve and dull ends were perfect for the job. For the most part, the bark peeled off in large sheets. We cut the logs into 20-foot (6-meter) lengths and peeled them. Bill skidded them to the landing, where we bucked them into 16-inch (40-centimeter) log-end lengths, our wall thickness. We split and stacked our harvest of cordwood near the building site for convenience.

First-floor Cordwood

It was not until August 1991, after four months of very hard work at the site and in the forest, that we began to lay up cordwood. Without the time constraint of the seasons, cordwood building would be a rather peaceful building method. A friend loaned us a gas-powered mortar mixer to help speed things along. The first time we fired it up, we knew it wasn't for us. Those of us who had experienced the noise and commotion of typical building sites wanted this to be a different kind of experience. We would gladly mix the mortar by hand in order to enjoy the quiet and the conversation.

Cordwood is an artistic collaboration — one person on each side of the wall, moving as one, choreographing log-ends into place. Each log is a small piece of the whole, creating a living wall. It is not an easy dance. There is give and take, the need for compromise, for leading, and for following. But the results are well worth the effort. In the hot sun of August, these odd, beautiful walls were taking shape. Our dream was beginning to manifest itself, and we were happy.

24.2: Lower storey construction of Pompanuck Round House, Cambridge, New York. Credit: John Carlson.

Next to the front door entrance, we laid up a beautiful 14-inch (36-centimeter) oval log end, with "Pompanuck 1991" neatly carved on it. This was our "corner log."

Progress was slow through August, for we were simultaneously building walls; finishing up the first story of the large, round Finnish stove at the building's center; and erecting the steel frame between the center and the perimeter walls.

September came, and we had completed most of our other tasks and could focus completely on cordwood. We had to. Fall was in the air and was beginning to grace some trees with color. We hired an extra person to help with the mixing and laying. We invited friends for a cordwood party to help get walls up fast. We worked from dawn to dusk, often pointing by car headlights.

The south-facing side of the structure has five, 6-foot-tall (1.8-meter-tall) window frames. We filled in the areas between these windows with cordwood, up to wooden plates to which our floor joists would be attached. Dark clouds bringing the first of the autumn rains showed themselves between days of extreme heat. We covered the completed walls with plastic when rain threatened and removed it on the sunny days to keep the mortar from baking and the cordwood from expanding by taking on too much moisture from condensation.

We could not keep the curing mortar from developing small cracks. Despite all our efforts, rainwater made its way via window frames into the curing masonry. And not only were these cracks developing, but the wet cordwood was also expanding and pushing on the window frames — in some cases forcing them out of square and deflecting the uprights over

½-inch (13 millimeters) inward. These frames were already cross-braced, but we added a series of horizontal braces to keep the openings at the proper width.

The mortar cracks were disconcerting. At first we thought we were overpointing the mortar, drawing moisture out and causing it to cure too quickly. So we modified our pointing techniques but still got cracking. Then we looked at our insulation mix, the sawdust/lime mix that separates the inner and outer mortar beds. We had been collecting planer shavings for years in our woodworking shop for this purpose. These shavings were made mostly from kiln-dried hardwoods. We hypothesized that the wood shavings acted like sponges, wicking moisture out of the mortar and causing it to dry too fast. We got a load of "green" sawdust from a local mill and used that instead. This seemed to give us better results, but still the cracking persisted.

We considered our mortar mix next. Our original mix was 9 parts sand, 3 parts Portland cement, 3 parts lime, and 3 parts soaked sawdust. We modified that by reducing the Portland cement to 2 parts and increasing the lime to 4 parts. Our intent was to create a slightly weaker but slower curing, more plastic mix. We even spent more time mixing to ensure greater plasticity.

By late September the rains had increased, wetting the site almost on a daily basis for a solid week. Those small, persistent cracks were being overshadowed by a much more pressing and more serious problem. The three, 6-foot-high by 3-foot-wide (1.8-meter-high by 0.3-meter-wide) wall sections we had built had begun to tip out of plumb. One had moved out 3 inches (7.6 centimeters) over its 6-foot (1.8-meter) height. The cause of this problem was inherent in the curve of the wall, the rather dense Japanese larch log-ends, and the rain. Despite all our efforts to keep the walls covered, the larch still took on enough water to expand measurably. (Rob and Jaki Roy had a similar problem with dry hardwood log-ends at Earthwood, described in Chapter 12 of Rob's *Complete Book of Cordwood Masonry Housebuilding*, Sterling, 1992).

Now, we were not happy. Once we saw that we could not brace the cordwood walls sufficiently to stop their outward migration, we needed to change tactics. We made a decision to stop building the walls and set about installing the radial floor joist system for the second floor. We felt that the joists, nailed in place and decked, would create enough resistance to the outward movement of the cordwood walls. We intended to complete the second floor deck and lay the rubber membrane that would eventually become the roof. We knew it would be better to build cordwood under the cover of this temporary roof. We attached lines around the tops of these leaning walls and connected them to a come-along mounted to the masonry stove. Where we could, we slowly winched the walls back to plumb. (Yes, it is possible to winch cordwood walls, but do not try this at home.) This worked for two of the three walls that were tipping out, but the third wall had to be demolished.

Once these errant walls were plumb, we installed all of the 2-inch-thick (5-centimeter-thick) wooden plates, attaching them to the tops of window and door lintels and on top of any cordwood walls that were complete to finished height. We then installed the ring of outer joists, those running from the steel ring to these outer walls. To ensure that the joists would hold the walls plumb until the decking was installed, we braced the joists from the steel frame to the base of the perimeter wall with long two-by-sixes. This created a trusslike system that held the joists against the outward thrust of the cordwood. The system held firm while we ran the decking on the joists from the outer surface of the wall to the steel, further stabilizing the structure. Once the decking was in place, we confidently removed the two-by-six bracing.

Wintering Over

We had designed the house roofing system to use the EPDM waterproofing membrane for an earth roof, so we could use this membrane as a temporary roofing for the first winter. The next year, we would roll it up and store it until we were ready to install it in its permanent place. We ordered the roofing in two 20-by-100-foot (6-by-30-meter) rolls. We used a backhoe to lift them onto the deck and we unrolled them to cover a layer of 2-inch (5-centimeter) Dow Styrofoam™ that would eventually insulate the roof. For this first winter, we overlapped the membrane three feet (about one meter) at the seams and made a temporary splice. We cut a hole in the center of the rubber to extend the Finnish wood stove flue above the deck, so we could heat the house over the winter. We had finally created a dry working area below — and we were happy again.

We returned to cordwood with the more plastic mortar and planned a cordwood party, inviting 12 friends to help. We seemed to get a reprieve from the early cold that we were having, and the first day of cordwood laying went well. Within two days we completed the walls, which were responding nicely to the additional bracing. Using cordwood, we started to lay up the snow blocking — the space in the wall between joists or rafters — above the plates that connected the tops of all our window frames. We got halfway through the snow blocks when severe cold arrived in early November. We covered the window frames and the 12 remaining snow blocks with plastic. We stapled 4-inch (10-centimeter) fiberglass insulation in the snow block areas to tighten the house a bit more and to create a place for mice to build their winter nests. Now everybody was happy.

Spring 1992

After our experience with wood expansion and walls leaning out (both functions of excessive moisture), we decided to build our second floor under cover, which meant getting our roof up before laying cordwood. Since we would have window or door frames in almost every other space between rafters, we had the beginnings of a post and beam system. We designed a structure that would support our windows and doors and the continuous wooden plate system that distributes the roof load to the walls.

Spring was beginning to show itself in small ways, which meant that we would have to face the inevitable: removing our temporary roof in order to build the second floor, thus exposing the structure to the elements once again. How would we keep the first floor walls, kitchen, and bathroom areas dry? It wasn't easy.

We erected the steel frame for the second story, and built a temporary roof frame over the center of the structure to pitch water to the perimeter. We covered the whole structure with two, 30-by-60-foot (9-by-18-meter) blue reinforced poly tarpaulins. This tarp roof evolved as the building progressed. When only the steel and temporary rafters were installed, the tarps covered this structure and ran to the floor at the outside edge of the house. We assembled all of our window and door frames under the cover of this tarp dome. As we stood the window, door, and post systems in place, we reconfigured the tarps to keep the structure covered as best they could. After a particularly long rainstorm, we returned to the site to find that the tarps had filled with water in large bulges. Two pools filled with thousands of gallons of water hung suspended over the second floor deck. These had to be siphoned out for hours with garden hoses before we could resume work.

With each new day of construction complete, we would leave time to conduct what we affectionately called "Tarp Wars." This consisted of a few (usually two) of us attending to the stretching and securing of these giant tarps to their best water-resistant position for the evening. Wind and rain sometimes added to the intensity of the experience.

In order to work under cover, we had to fabricate a post and beam system in the external walls, consisting of 4-by-8-inch posts attached to the sides of each window frame. This involved a lot of retrofit work and bracing.

Also, to provide rafter support at the center of the building, we needed to build the stone chimney on the second floor, the 5-foot-diameter (1.5-meter-diameter) upper part of our masonry stove. This was a labor of love, using stones we collected in our area, as well as rocks that were special to us, including some given by friends. Construction of the masonry stove is outside the scope of this book, but what we did followed quite closely the detailed example in Chapter 15 of Rob's *Complete Book of Cordwood Masonry Housebuilding* (Sterling, 1992). With the central stone mass and both the inner and outer frameworks completed, we were able to install the heavy 5-by-10-inch and 6-by-10-inch roof rafters, covered by 2-by-6-inch tongue and groove planking. Before the EPDM membrane was installed, we covered the entire deck with construction paper to protect the rubber against possible abrasion. At last, the roof membrane was set in place, the two seams glued and caulked, and the lead flashing cone for the masonry stove chimney installed. The Tarp Wars were over!

We have taken time to mention our travails with the hope that the reader can avoid similar problems. The authors agree that if we were to do a large round cordwood building again, we would start with a 16-sided post and beam framework (as described in Chapter 6).

Second Floor Cordwood

In mid-August, finally working under cover, we began laying up the second floor cordwood. Rather than building from floor to plate, we laid cordwood to a height of 24 inches (60 centimeters) all around the circumference. Each section was allowed to cure before more weight was placed on top. We believe that laying the cordwood masonry in runs or lifts in this way made the walls stronger and equalized pressure on the windows.

We also overcame the excessive mortar cracking we had experienced the year before. We discovered that the masonry sand we had been using in our mortar mix contained a high percentage of clay, creating a weaker bond between the Portland and sand particles. The clay may also have expanded somewhat.

It is always a good idea to check your source of sand before you buy it. A quick and telling test is to fill a clear glass canning jar with ¼ parts sand and ¾ parts water, shake it, and let it settle. Clay is lighter than sand and remains in suspension longer. Thus any clay — or silt — that is present will sit on top of the sand when the solution settles.

We changed our source of sand, so it was free of clay. We also changed our mix slightly by adding another shovelful of soaked sawdust. This retarded the drying time and set of the mortar. And we worked the mix longer for greater plasticity. These factors combined to help keep the cracking to a minimum.

Being under cover and more in control of the medium gave us greater flexibility to express ourselves artistically. We placed the log-ends with more care, and we custom cut small wedge-shaped "shorts" or "cheaters," which we tapped into the mortar to balance spacing. We experimented with elaborate bottle designs. On the first floor, these were small, carefully calculated patterns. But on the second floor, we made larger, more complicated patterns as well as free-form motifs — random points of color and light cascading between dark cordwood shapes. For snow blocking, we used a pattern of rounds that repeated in each bay. This facilitated the completion of this task while adding a pleasing visual element to the crown of the structure.

24.3: The new Pompanuck Community sauna. Note the ventanas naturales on the left, inspired by Hans Hebel of Chile. Credit: John Carlson.

During the course of this building season, we had a number of cordwood parties to which we invited friends from the surrounding area. These were highly productive and lots of fun. Many hands made the pointing go a lot faster. These were times that deepened friendships, and we think that cordwood is a perfect community-building material and technique. When all the dust cleared, we actually used about 10 of the 20 full cords of log-ends that we had prepared.

What can we say about living in this hand-built house? We love it! The rustic elegance of the cordwood, the wholeness of its circular shape, the light streaming in through windows, and the colorful bottle designs are all part of a magical living environment.

The process of building as a community with the larger local community's support has been exhilarating. This was not a project to tackle alone, and we are grateful to all of our supportive friends who came out to help.

In 2002, we completed a rather large, oval-shaped cordwood sauna, big enough for a dozen people to sauna together. We will continue to incorporate cordwood masonry into our structures as the community grows. Current plans are for John to build another round cordwood home for his family and for Scott to build a hybrid cordwood and straw bale home for his. The original building will then be a true community center and guesthouse.

CHAPTER 25

A New Home on an Old Foundation

Stephen and Christine Ketter-McDiarmid

BEFORE WE SHARE SOME OF THE EXPERIENCES of our "cordwood home journey," we should explain that when we started, neither of us had ever attempted anything even remotely like building a house. Not even a doghouse. Not even a birdhouse. We had the desire and the dream, and we hoped that with some tools (which we had yet to acquire) and with enough trips to the library (first carpentry books, then roofing, then plumbing, then electrical, etc.), we could learn what we would need to go for it.

Our cordwood home building project was a three-stage process. We began by dismantling an old, abandoned 19th century home, which took us about eight months' worth of weekends. We kept the original stone foundation — circa 1865 — which provided a full basement space. That foundation measured about 18 feet square (5.5 meters square), and we planned to use it to construct a small guesthouse as practice for a larger, round cordwood home on another part of our land.

As the process unfolded to the second stage, that of actually building anew, one idea led to another, and before long we were extending the stone foundation (welcome to masonry!) by expanding out the back and to one side (future kitchen bay). Then half of the front porch became the dining room, and soon the other half became part of the larger kitchen. The roof was then extended further over the front to provide a nice sitting porch. Suddenly the floor plan became large enough to qualify as our true home. Practice was going to have to be the real thing, and guests would have to bring a tent!

The construction method we chose incorporated a post and beam frame of locally sawn oak, to enable us to erect the roof before filling in the cordwood walls. The massive oak posts were 9-by-9-inch, with the tallest ones weighing over 600 pounds (270 kilograms). Erecting them by hand, with the help of friends, was a very challenging and satisfying part of the "homing" process. We had no crane, but we did have something even better: our good friend and neighbor Dave Savage. Dave (who runs a nearby sawmill) and his father cut all of our

25.1: Our post and beam frame. Credit: Ketter-McDiarmid.

beautiful oak timbers. The rich heritage of generations of Ozark loggers in Dave's family was evident as he handled beams by himself that two or three of the rest of us would struggle with. As we explained our unusual plans for our cordwood home, Dave was hooked! He continued working with us the rest of the way, earning the title of "Master Mortar Mixer" from Chris.

The most challenging placements were the 12-foot-long (3.6-meter-long), 6-by-8-inch rafters, hoisted by block and tackle up to their final resting place 18 feet (5.5 meters) above the floor. Perhaps they were the most satisfying, too, as they successfully wedged down into their angled positions — the "birds' mouths," we think they are called — calculated and chiseled out on the ground with no opportunity for any adjustments later! After the rafters came the roof, complete with a tongue and groove vaulted ceiling above the rafters and horizontal purlins (we were even beginning to learn the lingo!). Three opening skylights were installed, which make a huge difference to the amount of light upstairs and down.

Once roofed in, we began the exciting process of the cordwood infill. Perhaps exciting is not the most appropriate word. The floor, the timber framing, and the roof took us about seven months (mostly three-day weekends), but now there was no holding us back!

During the second stage, we changed our cordwood building medium from 12-inch (30-centimeter) oak log-ends, which refused to yield their bark, to 9-inch-thick (23-centimeter-thick) red cedar walls. The choice of the cedar presented itself when we suddenly realized that our land was blessed with rows and rows of old barbed-wire fencing on cedar posts, which we were planning on taking down anyway.

When we began to pull the old posts (some over 50 years old) we were in awe to find that the majority of them were still as solid as could be. When cut to 9-inch (23-centimeter) lengths with our miter saw, they showed their beautiful red hearts and revealed their special cedar aroma. And so our building material became a wonderful example of true recycling from our own piece of land. The curious passersby and neighbors, however, wondered what we could possibly be planning to do with that pile of old fence posts! Cutting fenceposts went quickly, and before we knew it we were mixing mortar and filling in our first section of wall.

We began by mixing in a wheelbarrow and used the Portland cement recipe we'd learned at one of Rob and Jaki Roy's three-day workshops in Wisconsin. The proportions were 9 parts sand, 3 parts lime, and 2 parts Portland cement. We had great results using a liquid cement retarder from our local concrete plant, at about one-quarter of a cup per batch. We later changed over to using a cement mixer but never changed the recipe, which worked beautifully.

There is one piece of advice we'd like to share. Try to buy all the mortar materials from the same supplier. We got Portland cement at three different places, and each had a slightly different color. The same is true for sand, where the texture can vary as well as the color.

25.2: Our 9-inch-thick (23-centimeter-thick) red cedar cordwood walls fill all the panels framed by the posts and beams. Credit: Ketter-McDiarmid.

The oak sawdust for our insulation — 12 parts sawdust to 1 part lime, as we'd learned in Wisconsin — came from Dave's sawmill and poured into the walls beautifully.

Chris became our expert pointer. (In fact she ended up pointing every single log-end in the entire house! She's just full of "pointers" now!) She also directed the placement of our cobalt-blue bottle-ends, which add a beautiful dimension to the east and west walls.

The cordwood process went quickly, in part because we were now devoting full-time, seven days a week, to the house. Three months after beginning the cordwood, we laid up our last cedar log-end. A change was as good as a rest, and we spent the next two weeks installing our red oak hardwood floor.

A wrought-iron spiral staircase finally replaced our temporary ladder to the loft bedroom, which also marked the end of the "cat-ramp" for our very patient but not ladder-capable pet. Internal plumbing and wiring commenced and continue to this day. (Perhaps we should have said that our house is a four-stage process!) Our finished home ended up just under 1,000 square feet (93 square meters), including the loft. The basement provides us with another 600 square feet (56 square meters) of storage/work space.

Well, we moved in — ready or not — and continue to complete the internal details like drywall, tiling, etc. (Chris, where's the drywall book?). But we love it more than we had imagined and would encourage anyone else to give it a whirl!

One important note: attending Rob and Jaki's workshop was invaluable in giving us the initial confidence to move forward. They gave us an opportunity to "feel" the mud as it should be mixed and to actually place log-ends into a real structure. (There is no substitute for experience.) Another great source of information and inspiration — and even some enlightening discussion — is Alan Stankevitz's Daycreek website, which covers all things cordwood. Both resources are listed in the Bibliography. Thanks to you all.

To prospective builders, we can only say, Enjoy the whole process. Enjoy this and every day!

Part Four
Go, Thee, and Do Likewise

26 • The Mortgage-Free Cordwood Home 183
   Rob Roy
27 • Cordwood and the Building Inspector 193
   Dr. Kris J. Dick and Professor A.M. Lansdown
28 • Cordwood and the Code 203
   Thomas M. Kwiatkowski
29 • Cordwood and the Code: A Case Study 207
   John Carlson and Scott Carrino
30 • Cordwood Code Issues: Strength and Insulation 213
   Rob Roy

CHAPTER 26

The Mortgage-Free Cordwood Home

Rob Roy

Here's a simple question: Do you want to own your home, or do you want the bank to own it for you? Kind of loaded, isn't it? Like the questions you get on questionnaires supplied by special interest groups. Who in their right mind wants to sign up for a lifetime of economic servitude? And yet people will wait in line at banks to do just that, especially in these times of relatively low mortgage rates.

In feudal Scotland — when the masses of people were known as "serfs" — three months of your labor went to the laird. In return, you got your land, shelter, defense, etc. Today in California, nearly half of people's after-tax income goes toward shelter alone. In New York, "tax freedom day" is late in May; the first five months of the working year go toward state and federal taxes, none of which contributes toward your shelter. If the medieval folks were serfs, what should we call ourselves today? You see, the questions get more difficult.

I realize that some folks will have to get a loan to proceed with even a low-cost housing project, such as a cordwood home. And many of the owner-builders in our cordwood masonry database have done so — usually taking out personal or construction loans. But I'm not the right guy to tell you how to do this. Jaki and I have always paid as we go, belonging to the old-fashioned school of thought expressed so well by Stewart Brand of *Whole Earth Catalog* fame: "Living below your means is a cheap way to be rich. It's the only way to be rich."

My last comments on lending institutions (before I tell you how to own your cordwood home) are that they frown on the unconventional: be it cordwood, underground, straw bale, cob, or whatever. They are hung up on such phrases as "track record" and "resale value." They want a drilled well, even when there is a perfectly adequate dug well on site. They want central heating, in an underground house that can never freeze. They want connection to commercial electricity — another good method of signing up for economic servitude. Well, one of the aims of this book is to give cordwood masonry more credibility and to make it easier to deal with building inspectors, insurance agents, and other paper people. Maybe even give it — hang on to your knees and things — "mainstream acceptability." What the heck, why should the benefits of cordwood masonry be limited to the radically sensible?

Okay, how do we avoid a mortgage and other insidious forms of enslavement? By adapting a few useful time-tested strategies. The first big step, which you've already taken, is to consider building a cordwood home, certainly one of the most economic building methods going. And for someone with suitable wood on site, cordwood's economy is even better. And, as a bonus, you get pleasing esthetics, energy efficiency, ease of construction, and ecological harmony.

The Grubstake

The word "grubstake" originally referred to money or provisions advanced to a prospector in return for a share of his findings. "Grub" was advanced for a "stake" in the claim. Nowadays, the term commonly refers to monies laid aside for the purchase of land or building materials.

And where do you get this grubstake? You may already have it. There is no better return on your investment than building a house, so any monies tied up in savings accounts — even time accounts that would incur a withdrawal penalty — are better put toward lowering shelter costs. Maybe you've got equity in the home you are now paying a mortgage on. Maybe you've got an outrageous car, the sale of which would finance a cordwood house. Really. You'd be surprised what your net worth might be.

But some of you will be starting with next to nothing. Roy's First Law of Empiric Economics is this, Work to save money, not to earn money to pay someone else to do what you can do yourself. A dollar saved is worth a whole lot more than a dollar earned, because we have to earn so darned many of them to save so precious few. Take advantage of genuine bargains on building materials during this pre-building period. Get into money-saving routines. Rent videos and fix a special dinner instead of going out to a restaurant and a movie. Give up smoking. (The cost savings here are compounded by lower lifetime health care savings.) Make beer instead of buying it. (The beer is better, too.) What? You don't have any of these vices? Too bad. You are, as Mark Twain said, "like a sinking ship with no freight to throw overboard." The point is that toughing it out for a year or two will often yield enough bucks to get to the land, where the real savings start.

The Land

Land can be expensive. Thoreau's $28 home on Walden Pond is all very impressive, but he built the house on Emerson's land, not a bad strategy. Maybe you have a relative that will give you a chunk of the "north forty."

If you are contemplating making a major lifestyle change, maybe you should consider moving to where the land is cheap. This is what Jaki and I did in 1975. And land is still cheap around here. And it's a great place to live. But I'm not trying to bring people into northern New York, only to let you know that there are places where there is good, relatively cheap land. To give you an idea: land in Vermont, just across Lake Champlain from us, is easily three times the cost of similar land here. Vermont's a great state, don't get me wrong, but geographically, land there is not much different to that in northern New York. But there's this mystique about Vermont. Rich people from New York City and Connecticut buy land near the ski areas or quaint villages and drive real estate values up.

It is beyond the scope of this chapter to go into all that must be considered in a land search. As pertains to economy, though, be aware of hidden costs down the line. If you have to have commercial electricity, what will it cost to bring the lines in? This can be a shocker. What will a well cost? Is a dug well a possibility or will you have to drill? How deep did the neighbors have to go? Are the soils conducive to an ordinary septic system, or will a very expensive system have to be built? Are alternative sanitary systems allowed? What about access — not only for you but also for concrete trucks, building materials, etc.? Is the land blessed with indigenous building materials? Do you have wood, stone, sand, topsoil (to grow food and for an earth roof)? These things can save you a fortune later on.

The Temporary Shelter (TS)

This is the first of the really great strategies you can use to avoid sub-serfdom, particularly if your land is still relatively close to your place of work. Build a temporary shelter (not necessarily a temporary structure) on your land and move into it, thus eliminating whatever shelter costs you are now paying, be it rent or mortgage. Now the savings mount up fast, as the formerly biggest part of your expenditures has been eliminated. For some, this might be $500 to $1,000 a month.

And what is the nature of this temporary shelter? Well, it should be small, quick and easy to build, and should employ the same building techniques that you plan to use in the main house later on, i.e., cordwood masonry. You see, there are several other advantages to this strategy besides eliminating shelter costs: Building experience is gained — a $500 mistake on the TS might save a $5,000 error on the main house. Knowledge about the land is gained while living on it — such practical information as where the sun rises and sets at different times of the year, and where the sensible access, well, and septic locations are. Maybe you're just not a builder. Better to learn that on a 300-square-foot (28-square-meter) shed than a 1,500-square-foot (140-square-meter) house. If you can't build the shed, don't start on the house.

26.1: Our 20-foot-diameter (6-meter-diameter) office would make an excellent temporary shelter. Credit: Jim Rhodes.

Finally, as the structure itself is not temporary (only its use as shelter), you will have an outbuilding for later use as a guesthouse, studio, workshop, sauna, or whatever. You could even incorporate the structure into the final house plans. The TS might become the master bedroom, for example.

Our 20-foot-diameter (6-meter-diameter) office at Earthwood would make a good TS, with 256 usable square feet (23.7 square meters) (*round* feet?) of space. Even in 2003, it should be possible to build it for $2,500 — three or four month's shelter cost for many renters.

Are you more of a square than a round person? (Neither sounds very complimentary!) No problem. The strategy will work just as well with a post and beam structure like *La Casita*, our guesthouse. In fact, square structures have the advantage of being easier to add-on to.

Keep It Small

Over 300 years ago, Thomas Fuller said, "Better one's house be too little one day than too big all the year after." This is true once again, after an unfortunate period of wasteful use of the planetary capital. But building small just for the sake of it serves no useful purpose, either. A family's space requirements fluctuate. Young couples with a small budget can live comfortably in a small house that would not be suited to a family with three teenaged children. A small house can be expanded, as need dictates and personal economy allows.

Although economy is the obvious reason for building small, it is not necessarily the most important one. The important reason for building small is to get the thing completed. Inexperienced builders, even those with plenty of money, should not tackle a house larger than 1,200 square feet (110 square meters), particularly a cordwood house, which is labor intensive. There is a very real danger that the place will never be completed. Or if it is, that the stress of building will irrevocably stress the marriage, too. Listen:

One lady, responding to our Cordwood Database Questionnaire said, "It's a beautiful home, visitors are thrilled by it, but it destroyed our relationship. We are presently trying to rebuild our marriage. I personally know two other couples who are going through similar problems after their cordwood home projects."

Another lady, in response to the question, What would it cost you to build this home today? replied, "A new husband!" Jaki and I experienced marital stress building the large Earthwood house, and this after having built two homes previously. So you have been fairly warned. In reality, cordwood masonry isn't any more stressful than other building styles. We know of more broken marriages where other forms of building were employed! Size of the home and frame of mind (karma) are more important concerns than technique. In fact, many cordwood builders find the masonry work itself to be quite therapeutic.

There are lots of questionable reasons why people think they need to have a big house, aside from bank propaganda and outmoded zoning regulations. Here are two biggies:

26.2: La Casita, a guesthouse at Earthwood, would make a good TS for a single person, while providing a lesson in simple post and beam framing.

- *The overreaction syndrome.* Jack and Jill (not their real names) have been cooped up in their little apartment for so long that all they can think of is, When we build our cordwood house, there's gonna be plenty of space! They've got lots of time to plan; paper and pencils are cheap. They finally get started on their 3,000-square-foot (280-square-meter) masterpiece. The possibilities from there, in descending order of probability are: 1. There is a great enthusiasm to begin with. After six months, money, energy, and patience run low, then run out. Jack and Jill split up. 2. After a while, J and J perceive that they've really bitten off too much. They move into a third of the place. Someday we'll finish the rest, they say. 3. They pull it off, as planned. I have only heard rumors of this scenario.
- *Bedroom mania.* The functions of a bedroom are to supply a peaceful venue for horizontal storage of the body, and as a catch-all, generally for clothes that are no longer used. You may think of other activities. But the bedrooms in most American homes could be divided in two, and each would still meet the need. Sure, other considerations come into the planning: building codes; an adjustment, perhaps, of the individual value system; planning a small bedroom to accommodate furniture. One thing is certain: the larger the bedroom (or house, for that matter), the more unnecessary clobber one accumulates.

How many bedrooms? Americans seem particularly concerned with the issue of privacy. Every kid has got to have his/her own bedroom, and then we throw in an extra one for the pot: the guest room, used 5 percent of the time or less. Our living room has a moderately convertible sofa, so it accommodates guests, too. That the sofa is not too comfortable has the side benefit that guests are less likely to overstay their welcome.

Keep It Simple

"Keep It Simple" should not be confused with "Keep It Small." A small house can be hopelessly complex, and a large house can be wonderfully simple. Here are some corollaries:

- *Keep to one style*. There is a style that suits your karma and pocketbook better than others. Once you find it, stick with it. If two house shapes are to be combined or intersected in some way, let there be a unifying force to the architecture: the use of cordwood masonry, for example, or a constancy of roofing material. A hodgepodge house always looks like a hodgepodge house.
- *Avoid difficult lines*. If you think they're tough to draw, wait until you try to build 'em! Keeping simple lines is of particular importance if you're inexperienced. Gambrel, hip, and valley additions (and dormer additions) should be avoided on the roof line, for example. Sunken living rooms, complex stairways, and split levels all add to the complexity — ergo, to the time and cost of building. Domes and polygons may have a strong appeal, but know that the finish work is long and tedious; furniture is designed on the premise that gravity runs perpendicular to the horizon; and it's unlikely that there will be local people experienced in these techniques to help when you get in trouble. If in doubt, build a model of the intended structure. "Cordwood Jack" Henstridge says, "If you can't build the model, don't try to build the house."

  Although octagons and hexagons can be built with cordwood masonry, the joins where walls meet can be tricky. It's quite difficult with stackwall corners, and post details have to be carefully planned (see Chapter 7). While not impossible — several have done it — my view is that if you are considering an octagon, it is easier and cheaper to go that extra little step to fully round. And if I were building a large round house again, I'd go with the 16-sided method developed by the Frasers (see Chapter 6) and repeated successfully by Jim Juczak (Chapter 14) and Alan Stankevitz (Chapter 15).

  Don't be afraid of round just because few people do it. Amongst the other building animals — not to mention so-called primitive people — practically no one does it any other way, precisely because round is simple, particularly with masonry

Impact of Perimeter Shape on Area

- Shape A, the circle. The choice of other building species. The most space per foot of perimeter.

- Shape B, the square. The most efficient rectilinear shape, seldom seen today.

- Shape C, the rectangle. The most common house shape today. Why?.

- Shape D, the 1950s mobile home. The longer and narrower we make it, the less space we enclose. We could build 59 feet (18 meters) long and 1 foot (0.3 meters) wide and have 59 square feet (5.4 square meters).

- Shape E, the architect gets involved. If those two "inner" walls had been left on the outside where they belong, we'd have the efficient Shape B. The roof is more complicated to build with Shape E and 225 square feet (21 square meters) are lost.

- Shape F, the "hockey rink." Many people are building cordwood homes like this, and not just in Canada. Still over 10 percent more space than the most efficient rectilinear shape (B), which almost no one builds. The roof is not complicated if a radial rafter system is employed for the half-circles. The radial rafter corresponding to the internal arrows on the diagram is also the first of any number of parallel rafters for the rectilinear section.

In fairness, it must be pointed out that if you enclose more space, you will spend more time and money on roofing and flooring. However, these components go faster than the labor-intensive cordwood walls. The important point here is that you can get any desired floor area by building less perimeter wall. Also, with less skin area, the home is more efficient to heat. Just thought I'd share this with you.

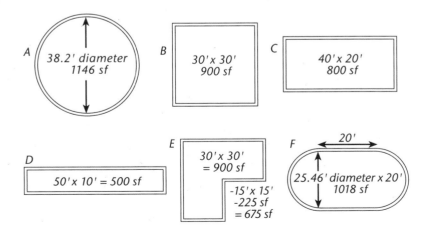

26.3: All six of these house shapes have a perimeter of 120 linear feet (36.5 meters). Look at the varying square footage figures.

units like cordwood. Keep in mind what birds, bees, and beavers know instinctively: a round house of a certain perimeter will enclose 27.3 percent more space than the most efficient rectilinear shape, the square. And most people don't even build square. They build a rectangle twice as long as it is wide, like we did at Log End Cottage. The free space gain of round over this rectilinear shape is better than 40 percent! So for the same amount of labor, materials, and money, the external walls of a round house enclose 40 percent more space than the rectangle. (See the sidebar "Impact of Perimeter Shape on Area.")

Now that's big savings! And if it saves you from building with hired money, the on-cost saving is compounded daily. And the round house is easier to heat because it has less skin area (heat loss) to enclose unit volume. I know, I know, domes are even better in this regard, but we're talking cordwood here. If you do decide to go round, I strongly advise the radial rafter system, as opposed to parallel rafters or pseudo-hip roof systems. Much easier to build. Think about it.

Okay, you've thought about it, and you — or your spouse — just can't make the jump to round. At least keep it square. And for heaven's sake avoid Ls, Ts, Us, and other projections, which further decrease the ratio of usable space per unit cost of materials and increases the time of labor (which is your time we're talking about here).

- *Avoid basements.* It is surprising how many people in North America continue to view a basement as a prerequisite to house construction. This is despite the fact that in a low-cost home, particularly a cordwood home, a basement will eat up a third of the building budget while providing low-quality space that gets used less than 10 percent of the time. Most typical basement functions, including heating systems, are best enclosed in the house proper. Pure and simple, basements are not cost effective, require familiarity with additional structural systems, and provide low-quality habitat for almost anything except large-scale mushroom propagation. If you are still not convinced, then I implore you to spend a little more money on insulation, waterproofing, ventilation, and natural light sources, and transform the basement into warm, dry, bright, airy earth-sheltered space, as we did at Earthwood.
- *Fit the floor plan to the structure, not the other way around.* Novice owner-builders commonly draw floor plans first, then try to design a structure to fit them, which often leads to complicated structural plans. My approach is to design a simple (therefore economic) structural plan, and then allow the floor plan to be somewhat shaped by the structure. Although a few compromises might need to be made, the end result is a structurally sound, easy to build, low-cost home. Earthwood makes use of this strategy. Internal rooms follow the lines of the main bearing girders, posts, rafters,

and joists. If internal walls just miss these members, this makes for nightmarish carpentry. Slows the project right down. And the finish work never gets done.

Use Recycled Materials

Recycled building materials are often better, cheaper, and have more character than new stuff. Using them is kinder to the planet, too. 'Nuff said.

Work Parties

Occasionally, it will be advantageous to throw a work party, particularly when it's time for the floor and footing pours, and again at rafter and roof work. Cordwood work parties aren't too effective, unless your volunteers are willing to donate a few days of work. It takes a couple of days to train people, during which production actually slows down. Also, you want to be very careful about quality. After all, you've got to live in this house and look at any shoddy workmanship every day.

And be organized on the big day, *before* the big day. I've seen owner-builders arrange for several friends (sometimes too many) to come over to help and end up playing hosts, serving up a case of beer while the crew stands around jawboning. Bummer. The owner-builder must be sure that all the required materials are ready the day before, that jobs are properly organized, that the workers will have — or be supplied with — the right tools for the job, and that there are no pesky little details that have to be attended to before work can begin. You'll get a week's work done in a day with organization. People come expecting and wanting to work. If they don't have a job to do, be sure that they'll start in on the beer, thinking, *Might as well make a party of it. I've blown this day coming over for nothing anyway.* Don't let this happen. Plan ahead. And involve everyone.

The Add-on House Strategy

Sometimes the temporary shelter, already discussed, will serve as a part of the completed house, either by plan or by evolution. However, get one part of the house completely finished before moving on to the next part. Living in a house under construction puts tremendous strain on a relationship. If you can retreat to a clean and uncluttered space, this refuge may prove invaluable on all fronts.

There are two different approaches to the add-on house strategy. One is to have some specific expansion plan in mind at the initial design stage. The other is to let the house grow organically as needs arise. Either approach will work, so tailor your strategy to your personality. If you have an analytical mind, you may be happier knowing that you are working toward some specific end. A more spontaneous individual might feel cramped by such a plan, preferring free creative rein throughout.

The add-on strategy is, essentially, a "build as you can afford" approach. Like the temporary shelter strategy, with which it is sometimes combined, it requires the ability to tough it out in less than the desired space for a while. Let's look at how it affects cordwood masonry in particular.

Of the three primary styles, the order or adaptability to the add-on strategy is: 1. post and beam with cordwood infilling; 2. stackwall corners; and 3. round or curved wall. Post and beam construction is a modular system, which is what we are looking for in a home with add-on potential. The stackwall corners method also has flat straight walls to add-on to, but there is the slight problem of knitting the new building into the existing stackwall corners. However this method has been used many times successfully. Round houses are extremely difficult to add-on to. If you absolutely must, follow the radial rafter lines outward and make a trapezoidal addition, such as the solar room at Earthwood. What really looks bad is to add a square room to a round house. And it's difficult to do.

One final technical point: it is easier to add-on to the gable ends than to the eaves of a square structure. Adding on to the eaves involves a shallow-pitched, shed-type roof, which can be very troublesome in the winter with heavy snow loads and potential ice-dam damage. Ceilings will be low in the addition unless the core unit had very high walls. It is easier, cleaner looking, and structurally superior to add-on to the gables.

I hope there's something in this chapter that will help you on your way to economic freedom, for it is my sincere belief that if you build your own mortgage-free cordwood home, you'll be more than halfway there.

(Author's Note: The strategies outlined in this chapter are fleshed out in my earlier book, Mortgage Free! Radical Strategies for Home Ownership, *(Chelsea Green, 1998).*

CHAPTER 27

Cordwood and the Building Inspector

Dr. Kris J. Dick, P.E. and Professor A.M. Lansdown

There is a need for methods of house construction that are different from standard frame or brick structures, methods which employ indigenous materials and skills. Examples of this need for decent housing abound for individuals in small, often remote and economically marginalized communities.

Over the years, the Northern Housing Committee at the University of Manitoba in Winnipeg has observed that for innovative housing to be acceptable by those who need it the most, it must meet some rather tough sociological criteria. Innovative housing techniques must be seen as acceptable by the dominant society, both in terms of style and technical merits. That is, they must incorporate standards equal to, or greater than, those of the conventional housing solutions that dominate the market. At the same time, however, these techniques must be much more accessible to the very users who need them the most — the people of the small communities mentioned above. Relatively few people that are involved in the self-help housing movement are aware that they are, *de facto*, on the cutting edge of regional and local economic development. This represents the real strengthening of North American communities, in spite of the mean-mindedness of the financially driven economies of the United States and Canada.

It is clear that any discussion of self-help housing that makes use of regional natural resources and "people power" (men and women working together) merits a full chapter of its own, but here we will concentrate on the construction techniques of cordwood masonry as they relate to presenting a proposed design to building inspectors.

Obtaining a building permit can be a relatively simple procedure. For some owner-builders, however, it is perhaps the most traumatic and angst-producing component of the entire building process. This chapter lays out the basics of building inspection, indicating the principles behind codes and permits. A study of the technical side of building reveals that there are six fundamental aspects associated with the approval of a building permit for a dwelling: structural safety, durability, energy management, moisture management, fire protection, and site location.

Structural Safety

Structural safety concentrates on foundation design and the structural strengths of walls, roofs, and floors. Foundation concerns normally focus on the basic load-carrying capacity of the foundation and the ability of the proposed design to resist heaving due to frost and moisture changes. If the proposed design addresses these issues directly and knowledgeably, then problems should not arise in the permit application.

Shortly after its inception in 1972, Manitoba's Northern Housing Committee realized that there were serious problems associated with foundations on structures in remote northern communities that the committee was studying. Many early log structures had been built directly on the ground, or "sill logs," allowing rising damp to rot out the logs differentially. As the house settled unevenly, air and heat management was lost. Serious distortions appeared in the structure. A majority of houses had concrete foundations, either in the form of full basements or as grade-beams. Rarely had expensive subsurface soil investigations been conducted prior to construction, and often the concrete work had not been done well. Consequently, concrete cracking — often massive — was the rule rather than the exception. A high water table, frost-susceptible soil, and permafrost all conspired to challenge a foundation's life. The committee recognized that subsurface investigations were impractical for all but the affluent and so decided to design a foundation that was strong enough for the worst soils likely to be encountered and one that was repairable. Some of the heaviest loads in any region are the axle loads of trains, so a railway-inspired solution was adopted. In this way, strength, frost, and moisture were addressed. Details of this foundation system appear in *Stackwall: How to Build It* (A and K Technical Service, 1995) (see the Bibliography).

Laboratory tests on stackwall (cordwood masonry) segments in the 1970s indicated that they could carry about 30,000 to 40,000 pounds per lineal foot (437 to 583 kiloNewtons per meter) which turned out to be 20 times the design load for a single-story building in regions with the heaviest snow loads. We concluded that, for one- and two-story houses, wall strength was not a problem with cordwood masonry. Conventional floors and roof systems, following manuals of good practice, have kept our house builders out of trouble on this front.

Durability

Durability of a cordwood structure is mostly a matter of moisture movement and foundation design. The question really is, How fast is the strength and integrity of the house compromised through rot and differential settlement? Durability can only be proven with time. Old structures, however, indicate that stackwall buildings are at least as durable as heavy timber structures. Known ages of some structures are: Manitoba — 50 to 100 years with poplar; Ottawa and St. Lawrence valleys — 100 to 200 years in various species. The oldest we have heard of is a monastery in northern Greece built about 800–900 AD and still in use. By way of comparison, many conventional frame structures in remote communities in northern Canada are in serious trouble within ten years.

Energy Management

The management of energy — via insulation standards; size, nature, and position of windows and doors; and management of fresh air — influences the annual cost of operating the home. Energy code standards are set in an effort to minimize energy costs in a home; hence the specifications for minimum R-values in walls and attics, for example.

In its earliest work, the Northern Housing Committee explored the desired optimum R-values for walls. For the price of fuel in remote communities and typical frame construction costs, it was clear that the old R-7, or even the (then-proposed) R-14 standards were far from optimum economically. We sought a minimum insulation value of around R-20. Using poplar log-ends, the 24-inch (60-centimeter) insulated cordwood wall we recommend has an insulation value of about R-20. In many cases, however, the actual R-value seems to exceed R-20 by a large margin. Caulking of air gaps, where drying and shrinkage have taken place, is a necessary part of construction. Caulking should be done no sooner than one year after construction to allow for full drying of the whole wall to take place. Otherwise, surplus air infiltration can ruin the effective R-value.

It is important to recognize the effect of thermal mass on heating in a building, since a large mass of material with a high unit heat capacity acts as a heat storage reservoir. A heavy building often turns out to be easier to heat (and stays cooler in summer) than a light frame structure having identical R-values for insulation. Experiments with full-scale stackwall buildings have indicated effective R-values considerably higher than calculated R-values.

Moisture Management

Moisture management is perhaps the least well understood of the inspection criteria. Witness the almost slavish dependence on poorly installed vapor barriers in many structures, especially in northern communities. Consider that a normally functioning residence is constantly generating water every day — from perspiration and exhalation of its residents, to cooking, bathing, and clothes washing. In the summer or in a warm climate, the moisture is not a problem since windows and doors are often open, allowing for the escape of excess moisture. In a northern climate, however, to conserve energy during winter, the house is often rather tightly sealed. Any generated moisture has to really struggle to get to the outside atmosphere.

A substantial portion of the excess moisture will try to escape through the walls. A framed wall filled with insulation provides a particular challenge. Without a vapor barrier, the situation is as shown in Image 27.1(a). Moisture passes through the wall and even though the whole section is permeable to vapor, trouble strikes. As moisture passes through the insulation the temperature drops to the dew point (point of condensation) where it condenses to liquid in the insulation. The soggy insulation loses R-value, and this retained moisture can actually rot the studs and outer wall. Image 27.1(b) illustrates what a good vapor barrier should do. It sends the moisture back into the room. Your windows may be a mess of frozen and melting ice, but your walls will be safe! Image 27.1(c) illustrates the usual situation — a good vapor barrier with a few pinholes provided by tears, punctures, nail holes, staple holes, and gaps around electrical fittings. Most vapor is sealed in but some

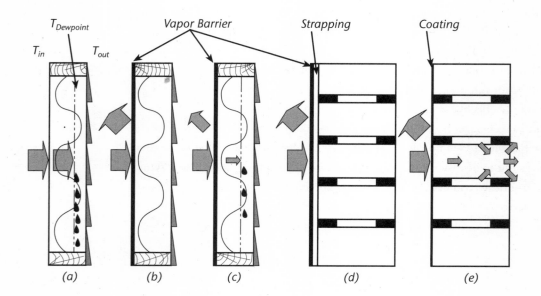

27.1: Vapor management in walls. Credit: Kris Dick.

escapes and condenses in the insulation. Tests in Sweden have indicated that a pinhole in a vapor barrier is enough to allow 12 to 25 pounds (5 to 10 kilograms) of ice to form in a wall over one winter! Only one drop of water is needed for the dry rot bugs to get started, and once they are off and munching your walls, they'll give off water to keep the process going!

Images 27.1(d) and 27.1(e) indicate cordwood masonry walls. Image 27.1(d) shows a stackwall section with strapping and vapor barrier on the inside. This ruins the esthetic effect of the wall style but may have to be done to meet some regulations. Image 27.1(e) illustrates perhaps the optimum solution — exposed cordwood masonry with a waterproof coating on the inside of the logs, acting like a leaky vapor barrier. The coating should be something that will slow down moisture movement along the end-grain. The coating can be urethane, PEG (polyethylene glycol) or even paste wax. The key concept here is that there is a route for the vapor to the outside. The whole wall acts as a vapor dump, in effect. Our experience here has been that even unprotected walls are naturally drying and do not pose a moisture management problem.

Fire Protection

Advice received by the Northern Housing Committee indicates that fire protection aspects can be considered under the headings of: smoke generation; flame spread; fire penetration; egress; and electrical distribution details.

- *Smoke generation.* A major issue in dealing with potential fire in a house is the volume and quality of smoke generated. Usually, wood-based smokes are less hazardous than those generated from plastics, clothing, carpeting, and furniture stuffing. With cordwood masonry, so little flammable material is presented in the wall that the generation of smoke in a fire is minimal.
- *Flame Spread.* A major question in residences is, How fast can flames spread along a surface? In Canada, ratings are measured against asbestos panels (Rating = 0) and Red Oak paneling (Rating = 150). Cordwood masonry, because of the mortar breaks, has a flame spread close to zero, if left exposed. The mass of the mortar draws heat from the fire.
- *Fire Penetration.* Penetration of fire through walls is a measure of fire protection from room to room or from adjacent buildings, and it is measured in minutes. Only a short time is required to exit a single-family dwelling, while two to four hours is a typical requirement for a commercial building. Cordwood masonry, because of its mass and mortar content, is rated at two to six hours. The senior author was advised

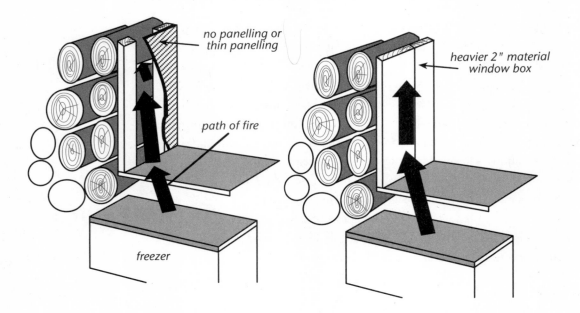

27.2: Fire penetration examples. Credit: Kris Dick.

by the National Research Council (NRC) Fire Division that tests were clearly not necessary, as stackwall met fire standards for tall adjacent urban structures. An interesting case is illustrated in Image 27.2. A stackwall store near Traverse Bay, Manitoba had a propane-fired freezer that exploded, causing a fire. In this case, a weak point was found by the fire as shown in Image 27.2(a). The windows had not been boxed in with 2-inch lumber (Rating = 65 minutes) but had been formed with almost no fire barrier to the insulation. The short-duration fire caught hold of the insulation. This could have been prevented by the detail in Image 27.2(b), recommended by virtually all of the cordwood masonry author-builders.

The fascinating feature of this fire was that it took two days for the building to be destroyed. During that time, with the guidance of the insurance adjuster, all furniture and fittings were removed at leisure. The committee was told that it was the best fire the insurance company had ever witnessed. Insurance rates for stackwall dropped like a brick after this fire. The senior author, an ex-firefighter, wondered what would have happened if the owner and the insurance adjuster had put the fire out by sprinkling water into the insulation from above in the two days allotted to them!

- *Egress*. An important point in a house design is an understanding of how any occupant could escape from a fire in any part of the structure. Always have at least two escape routes from any point in the home. The simple routes of egress are, of course, doorways, but do not discount windows. A word of warning here: high,

horizontal windows do not count as egress. We have lost more Canadian Native children in our north to the false pretence that these windows are legitimate egress than to any other building cause. They don't satisfy fire code, anyway, and for good reason. Don't use them. At the design stage, pay attention to the code with respect to the size and type of windows that provide safe egress, and how close the windowsill must be to the floor.
- *Electrical distribution details.* Fire caused by electrical faults is not an uncommon problem. The Committee has found that for a very small incremental cost, the builder can change an electrical firetrap into a very safe abode. We recommend that all wiring be done with protected armored cable such as flex cable or conduit. We feel that Chapter 10 in this book, "Electric Wiring in Cordwood Masonry," covers this subject very well and with a view to the requirements of the National Electrical Code in the United States.

Site Location

Site location matters tend to be simple but significant, nonetheless. They include issues such as height clearances, boundary clearances, and proper access concerns. If each of these has been addressed, there should be no inspection problem. Water supply and waste disposal are somewhat more formal requirements, and you will need to follow health code regulations in your jurisdiction: town, county, state, or province. Your water supply must be safe from natural problems, from your neighbors, and from yourself. Cordwood masonry does not present any extra issues not present in the correct siting of any home.

Having a good understanding of the issues discussed above is very important. Now it is time to get that building permit.

The Approval Trail

Those two little words, "building permit," can cause acid to grip the stomach. This need not be the case, if you keep three basic things in mind when embarking on the approval process for cordwood construction:

1. In most cases, the building inspector is likely unfamiliar with the cordwood concept.
2. By virtue of the lines of accountability within the local government structure, the building inspector may adopt a bureaucratic approach. It may not be possible to discuss the principles of this construction style but only where it fits the local rules.

3. In spite of local codes and restrictions, most building inspectors are interested in this construction style. Their main concern is that good building practice is maintained throughout design and construction.

The applicant must have a thorough understanding of the building technique. If you are not confident that you can explain details during that first visit to the inspector's office, take someone with you who can. The initial step in the approval process must be a constructive one, helping the process to move forward and not stalling it. Preparation is the key. Foundation design and vapor management, for example, are two primary concerns of many inspectors. Preparation of the site, drainage, foundation materials, behavior of the foundation under load, and the migration of moisture in the structure would be issues the applicant should be intimately familiar with and able to explain. If possible, make copies of the documentation related to cordwood masonry (such as these chapters in Part Four) and leave them with the inspector to go over at leisure, outside of a busy office environment. Or leave him the whole book.

Feedback from inspectors should be welcomed, not feared. Based on their wide experience, they can provide valuable input that may enhance the performance of the structure. Patience and humor are valuable life skills. Take them with you into the approval process, and the experience should be virtually painless.

Three Brief Scenarios

Approach is everything. The following are three possible ways to get a building permit. It is left to the reader to choose the one most suitable to them. These scenarios may or may not be true. Someone had a friend whose cousin …

1. *I need a rubber stamp. I'm on a mission.*
   Applicant: (Monday, March 20) "Good morning. I'm building a cordwood house five miles south of town. Got materials coming this morning. If you'd just stamp this drawing, I can be out of your hair in five minutes."
   Inspector: "Oh yes. Do you have plans? I'm not familiar with that style of building. Could you leave me a copy to go over for a couple of days?"
   Applicant: "Ah, don't worry. Here's a sketch. I talked with a guy who built one of these. He had an engineer do the design. No big deal. Can I have the permit?"
   Inspector: "Well, I'm pretty busy at the moment, got five major construction projects in town. Contractor's waiting for me out on the site. Let's see, it's March.

Best come back and see me in, say, late August. Probably be a good idea to phone first."

2. *Not ready yet?*
   Applicant: "Hi. I was in a couple of days ago and left you two books, five articles, six drawings, and a video about this cordwood house I'm building. Permit ready?"
   Inspector: (*Groan!*)

3. *Let's work on this together.*
   Applicant: (January) "Good morning. My name is ___ and I just bought the old Miller place south of town. I'd like to know the process for getting a building permit to construct a cordwood house."
   Inspector: "Cordwood? Never heard of it."
   Applicant: "Well, I brought some information and design details to leave with you. It won't be building season for another three months, so no big rush. When would be a good time to meet on this again?"
   Inspector: "A couple of weeks should give me time to go through this. Thanks."

Conclusion

We've discussed some of the design details related to cordwood construction that would likely be of interest to your local building authority. The actual building of a cordwood home can be undertaken by a few people or by many, as the case studies in this book have shown. If the fundamental design principles are adhered to, the end result should be a functional structure that will demonstrate an economically sustainable option for housing, and one which is kind to the planet at the same time.

Three things to remember: Have a thorough understanding of the principles behind the structure you are about to build. Formulate reasonable and realistic time lines for your project. And cultivate patience and humor. These strategies will not only help in the approval phase but will also help you to see the project through to a successful conclusion.

CHAPTER 28

Cordwood and the Code

Thomas M. Kwiatkowski

I WILL COVER CORDWOOD AND THE CODE topic with reference to the *N.Y.S. Compilation of Codes, Rules, and Regulations*. Different states use different codes. However, there are more similarities than differences among the various codes. In addition, your local jurisdiction may have its own zoning, land use, or housing standards.

The purpose of building codes is not to keep you from building the type of structure you want. The codes are there to help assure that the structure will not impede on your health, safety, and welfare. Of all structures, owner-occupied single family homes have the least restrictions placed on them by the code.

You will need to get a building permit. There is no way around this. To make this process the least painful and to ensure speedy results, I strongly suggest that you visit your local building inspector before you commit to a design or specific plans for your structure. During this first meeting, you will be able to get a feel as to how receptive the building inspector is toward alternative construction.

If the building inspector is unfamiliar with cordwood construction, you will probably be met with some amount of skepticism. So have a talk with the official. Explain cordwood construction; show a rough sketch of your project; and provide background materials, such as books, articles, photos, even videos. Also, at this time, find out all local land use regulations. The building inspector will be able to tell you requirements such as light (required window fenestration for various rooms), ventilation, size of

28.1: Tom Kwiatkowski's 12-sided cordwood home near Plattsburgh, New York.

structural members, snow and wind loads, energy code requirements, etc. Perhaps you are in a hurricane, tornado or earthquake zone, with special codes in force for these.

Now for the first (and what I consider the largest) stumbling block. Your plans for the structure will have to be signed and sealed by an architect or engineer that is licensed by the state or province where you live. Some states may not have this requirement. Others, like New York, may only require stamped plans for houses over 1,500 square feet (140 square meters) or some other size. In New York, an engineer does not have to be a structural engineer. He or she can be an electrical, industrial, environmental, or any other kind, as long as they are licensed by the state. This is a loophole you might be able to work to your advantage. Also, the architect or engineer does not have to be the one to draw up the plans. You or someone else can draw the blueprints, then pay an architect or engineer to review, sign, and stamp them. Shop around for prices, which vary a great deal.

As far as cordwood construction, getting your plans signed and stamped may be the most difficult steps in the process. This is due to a shortage of engineering tests available on cordwood construction. However, if you choose the post and beam style with cordwood infilling, all of the loads and stresses will be based on the framework. Also, there are companies out there that provide blueprints for post and beam structures already signed and stamped.

There are many discretionary aspects to the code, and final permit approval rests at the local level. Therefore, I strongly suggest that the better the working relationship between you and your local official, the easier it will be to complete your project. You will probably be building in a rural area, and in those areas the building inspector is usually a part-time position, so you will not be scrutinized or inspected as if you were in a more suburban area.

As for the discretionary aspects, one example is electricity, which may or may not be required for an owner-occupied single-family home, at the discretion of the local jurisdiction. The same is true for the required plumbing fixtures and hot water. Hook-up to local commercial power, however, must be approved by the local power company, and they will most likely require adherence to the National Electrical Code (NEC) and might even have other local service entrance requirements.

(Editor's Note: The rules for how many feet of line will be supplied for free can vary widely from place to place, too, so if your site is remote, it is good to check up on this in advance with the local power company. If hook-up costs approach $10,000, alternative energy may be a viable economic option. Incidentally, the NEC also has regulations and codes on the installation of photovoltaic panels.)

Another example of a discretionary situation is the use of unmarked structural lumber, such as locally sawn, rough-cut timbers. If you and your local official have an adversarial relationship, he or she could require certification of the unmarked structural lumber by a structural engineer. This can be expensive.

As for heating, you are allowed to have a wood stove as the sole heat source. By using a wood stove only, you will be exempt from the N.Y.S. energy codes. But, if you are hooked up to electricity, and if you have over a 100-amp service, it is assumed that you will be using electric heat, even if it is not installed at the time. However, the local authority may approve a larger service if you show that the larger service is necessary for purposes other than electric heat. Once again, if you and the local authority don't get along, he or she can require you to provide an engineer's report stating that the wood stove can maintain a temperature of 68 degrees Fahrenheit (20 degrees Celsius) in all habitable rooms (measured at a point 5 feet [1.5 meters] high and 2 feet [0.6 meters] away from any exterior wall).

If you plan on using an electric or fossil fuel heating system, you will have to meet the energy codes. Do not let this deter you. The required R-value codes for the building envelope are there for your own comfort. A double cordwood wall with insulation between would certainly meet any state's energy code. (See Chapter 4, The Double Wall Technique.) Otherwise, you may have to convince the code enforcement officer that your cordwood walls meet the required R-value code.

(Editor's Note: Work has been done, and is being done, on R-values for cordwood walls. See Chapters 27 and 30.)

I suggest that you should install a heating system even if you do not plan on using it. It is very rare for a person or family to keep one residence for the rest of their lives. When I went to sell my cordwood house, which lacked central heat, no bank would hold a mortgage on it. I was fortunate that someone came along with enough cash to buy it outright.

I hope that you are not discouraged or deterred from using cordwood due to fear of building codes. Most building inspectors will try to help you in any way they can. Do not try to sneak something by them. Remember that they are actually there to help you.

(Editor's Note: Tom Kwiatkowski was a Codes Enforcement Official registered with the State of New York and worked for the City of Plattsburgh in this capacity. He designed and built a cordwood home in 1979 and lived in there with his wife Helen for 13 years. Tom died in 1997.)

CHAPTER 29

Cordwood and the Code: A Case Study

John Carlson and Scott Carrino

TO A LARGE DEGREE, uniform building codes were created to protect homebuyers from contractors who may otherwise use substandard materials and/or unsafe building practices. However, for the owner-builder interested in experimenting with building practices and materials that are not defined by these standards and specifications, like cordwood construction, the rules and regulations can create obstacles. If the builder does not take the proper precautions to get a project going smoothly, dealing with codes can be a frustrating and costly experience.

The owner-builder of a cordwood structure should work closely with code officials. If a builder is working in an area where cordwood has not been used before, it is important to begin this relationship well before drawing final plans or purchasing materials. Keep in mind that it is not the job of code enforcement officers to stop people from building what they want to build, but to assure that safe materials and practices are used.

Do Your Homework

Preliminary plans should be drawn and a support package created that graphically illustrates the details of cordwood construction. Except in rare instances, code officials will have little or no information about this building method. Use books like this one and videos showing existing cordwood buildings, as well as instructional videos. (The Bibliography has a thorough list of supporting materials that will help give code officials the understanding and confidence they need to approve the project.)

Bringing officials in on the ground floor of the project should help build a good rapport that will last through the completion of the building. Also, by helping to develop their knowledge base and interest, you'll help advance the use of this sustainable, esthetic and cost-effective building method into mainstream practice.

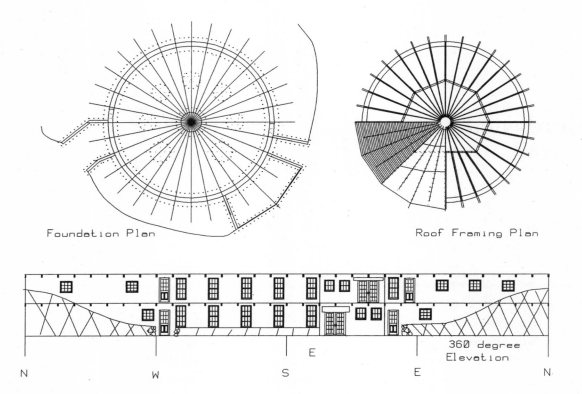

29.1: 360-degree elevation, Pompanuck Farm Community Center in Cambridge, New York.

With regard to stress load calculations, it is prudent to consult an architect or engineer to check the numbers. Mistakes could lead to structural failures that could put you or your family at grave risk. New York State for instance, requires plans signed by a licensed architect or engineer for homes larger than 1,500 square feet (140 square meters). This requirement can add considerably to the cost and complexity of a project and might be reason enough to decide to keep the project under this size. In the case of our 3,600-square-foot (335-square-meter) community center, we had to have an architect's stamp on our plans to get a building permit. We live in a rural area with no zoning regulations. Building permits here are issued under the jurisdiction of the Washington County Department of Code Enforcement.

In suburban areas, a builder may have to contend with even more rules and regulations. In other words, all work needs to conform to local, state, and federal codes. But where these codes conflict, the most stringent applies.

When an architect's or engineer's stamp is required, code officials are put more at ease, because the burden of responsibility for the safety of the structure is transferred to a professional. Apply the same process with an architect or engineer as with the code people:

show supporting documents and let them know what goes into a cordwood structure. If the builder draws the plans, the architect will use them to run calculations of the structural elements (floor joists, rafters, etc.) to see if they will safely support the structure. The architect who checked our plans came up with suggestions that saved money and time and avoided some problems with code regulations. For example, he designed a structural hinge system for our roof rafters. This significantly reduced the rafter dimensions and, therefore, the cost and weight of our roof structure. He also suggested that we use a steel frame to support the mid-span of our radial joists and rafters, again reducing the size and weight of these elements.

A Couple of Concerns

In areas of his expertise, our architect's suggestions improved our design. However, his lack of experience with cordwood proved to be a challenge to us. The cordwood wall system fell under scrutiny for two reasons. First, it combined elements (wood and mortar) which, according to standard building practices, are not to be mixed. Code requires wood in contact with concrete to be pressure treated or otherwise preserved from the possibility of decay from trapped moisture. After being somewhat convinced by our supporting materials that this marriage of wood and mortar would work, he suggested that we might run into problems with code officials by referring to "cordwood masonry" on our plans. We dropped the word "masonry" and, from that point on, used "cordwood" to define the wall system. Because of the unique way that cordwood combines wood and mortar — a log-end's ability to "breathe" and transpire moisture along end-grain — this breach of code did not alarm the officials.

Second, the architect was concerned that there were no independent laboratory test data on the compression strength of cordwood walls. There was nothing in the books that he could fall back on to defend his decision to approve cordwood construction. Our architect's main question (and worry) was, At what point will a cordwood wall system fail? There was a debate over which element — wood or mortar — he would base his compression calculations on. He chose the wood, even though we used a species (Japanese Larch) that did not show up on his charts. It took much discussion to have him believe, in theory, that cordwood walls were strong. (The compression strength of cordwood mortar is discussed in Chapter 30.)

In 1991, we completed the first floor of our two-story structure. As we rested for the winter, the architect began to worry about the structural integrity of our walls. At some point, he lost his nerve and informed us that he was going to withdraw his stamp from the project. We considered the time and expense of hiring another architect to get our plans stamped again and found this to be prohibitive. Instead, we decided to meet with him to seek a solution we could all accept.

After much discussion, we were required to reinforce the downstairs window openings to help take the load of the joists above. To accomplish this, we agreed to glue and screw a two-by-ten vertically to the inside of each window jamb. For the second floor, he gladly accepted a post and beam system we had designed to support the roof (so we could build the cordwood walls under cover). These two changes satisfied his concerns.

By putting his stamp on our drawings, he became liable for the structural soundness of the building; if it failed, the responsibility could fall on him. Consequently, he took many precautions to cover his assets and ensure that any mistakes that we might make would not come back to haunt him. On the blueprint page of general specifications, he included notes on lumber grading, the treatment of wood in contact with standing moisture, and that gravity loads must be carried in a straight line to the foundation, to name a few. He also included a note saying that, "this building system must be complete for stability; the builder must provide temporary bracing where necessary; the architect is not responsible for project safety."

Energy Code

Another area that caused us confusion and some consternation, was the new (as of 1991) New York State Energy Conservation Code, which requires new construction to meet certain energy use guidelines. There are two ways to comply with these codes. One is to rate a structure based on its components, such as windows, walls, doors, etc. The second is to rate the energy performance of the structure as a whole. This means that if one or more of a structure's components do not comply with the component standards, other more energy-efficient features might overcome that deficit. For instance, if an electric heating system is designed into a structure with triple-glazed windows, then the super-insulating quality of these windows might compensate for the wastefulness of the electric heating elements.

Initially we were excited at the prospect of having our cordwood structure evaluated for its energy conservation properties. We were pleased that the government was doing something to stem the waste of energy resources and saw this as an opportunity to bring cordwood a little closer to mainstream acceptability. However, these codes were not designed with alternative building techniques in mind. In our case, we knew our structure would be energy efficient, despite the fact that we were planning to use recycled, single-glazed sash with detachable storm windows instead of double- or triple-paned high R-value windows.

It caused us great concern when we were told to design and submit a second set of plans showing our structure built using traditional stick frame construction, as well as an analysis comparing it with the cordwood structure. This would have cost us a prohibitive amount of time and money. After further study of the regulations, we discovered that a structure not using fossil fuels for space heating and not installing more than 100 amps of electric service is exempt from the energy conservation code. We hope that in the future the use of recycled building elements, as well as the hidden costs to the environment of extracting raw materials and the manufacturing and transportation of new materials, will be considered as part of this energy conservation equation.

Attitude

It is important to keep a good attitude when faced with the inevitable frustrations of code specifics and how they relate to the building of a structure. For instance, after excavating, forming, and pouring our footings, the architect said that we needed to add 4 inches (10 centimeters) of washed stone before pouring the slab. This meant that we had to manually remove four inches of native gravel over the entire 48-foot-diameter (15-meter-diameter) area so we could add it back as stone. This would not have been a big problem if it had been done during the excavation with a backhoe or bulldozer. Instead, in the 90-degree heat of summer, we spent many days moving much material, for the sake of the code.

After the architect or engineer stamps the building plans, they are sent to the code enforcement authorities for final approval. When a licensed professional is involved in plan development and structural analysis, there is little reason to expect the code people will find fault. In our case, we had neglected to include a direct-wired smoke alarm. Adding it to the plan was the last detail needed to obtain our building permit.

Once construction begins, it is important to keep a clean (safe) building site and to notify code enforcement officers promptly at each phase of the process that requires an inspection. These include: footings before pouring concrete; foundation before backfilling; septic system; framing (for cordwood builders this would be wall construction); electrical; plumbing; and a final inspection before being issued a certificate of occupancy. Electrical placement in cordwood walls needs special attention where inspection is concerned. We ran plastic conduit along the floor and up into boxes aligned with what was to be the interior plane of the wall. Before beginning to build the cordwood walls that would cover these conduits, we requested an additional inspection. (Note: the electrical inspector in our county works independently of the state and county code enforcement agencies and requires an additional fee.) We found it prudent to keep a photographic record of details to help refresh memories in case misunderstandings developed after inspections of elements that became buried by walls.

By being conscientious about details and quality of construction, we had an easier time receiving approval than if our work had been sloppy. Following clearly conceived plans are a builder's best insurance against excessive scrutiny from building authorities, saving much time, money, and frustration. Inspectors are more willing to work with a builder who does not treat them as the enemy. It is an inspector's job to be helpful to the building process. If there are doubts about any step in that process, it is better to call and seek advice than it is to move ahead. Failing an inspection and having to redo hard work is a builder's nightmare.

(Editor's Note: In June of 2002, the first meeting of a new Cordwood Builders Association was held in Wisconsin, and by mid-2003, it is hoped that approved testing results on such things as strength characteristics of cordwood and insulation values will be available. Until the Association has a permanent address, correspondence will be handled through Earthwood Building School, 366 Murtagh Hill Road, West Chazy, NY 2992. Find us on the Web at <www.cordwoodmasonry.com>. You can keep up with latest developments through the cordwood masonry forum at <www.daycreek.com>.)

CHAPTER 30

Cordwood Code Issues: Strength and Insulation

Rob Roy

I HOPE THAT THE PREVIOUS THREE CHAPTERS will help prospective cordwood owner-builders through the permitting process. In this chapter, I will share some insight into the specific code issue of compression strength and insulation.

Certified Compression Tests for Cordwood Mortar

Just before the 1994 Continental Cordwood Conference, Paul Agnew of Cameron Geotechnical in Morrisonville, New York performed compression tests on six test cylinders of cordwood masonry mortar. The reader must keep in mind that while the tests were conducted according to standard New York State approved testing procedure, the words of this chapter are my own, not Paul's. The figures on page 214, however, are Paul's certified test results. I am not a licensed engineer nor do I play one on TV. I do have considerable experience with cordwood structures, though.

Code enforcement officers have little or no problem with the compression strength of wood. They regularly approve conventional horizontal-log structures all the time. The second component of a cordwood wall is the mortar, so Paul conducted several core tests on that, using the same standard test cylinders (6 inches diameter by 12 inches deep [15 centimeters in diameter by 30 centimeters deep) that he employs when testing the compression strength of concrete. All of the cylinders were filled and tamped on May 28, 1994.

Cylinders 1 and 6 were filled with mortar made in a wheelbarrow with the following ingredients, by volume: 9 parts sand, 3 parts soaked softwood sawdust, 3 parts Type S hydrated lime, and 2 parts Type I Portland cement. Cylinder 1 was tested to failure one week later. Cylinder 6 was tested to failure after 30 days.

Cylinders 2 and 4 were filled about 5¼ inches (13 centimeters) with the same mortar used with Cylinders 1 and 6. Then a wooden insert, a 6-inch-diameter (15-centimeter-diameter) disk cut from a pressure-treated two-by-six, was set into the mortar. Six roofing nails were nailed into each side of the wooden cylinder, but were left extending ½-inch (1 centimeter) proud of the wood, in order to "grab" the mortar. This simulates the same detail used on wooden plates used to distribute the concentrated load of rafters onto a cordwood masonry wall. The cylinder was then topped up with another 5¼ inches of mortar. Paul tamped all samples a certain number of times with a particular tamper, as he had been trained.

Cylinders 3 and 5 were filled with mortar made in a wheelbarrow with the following ingredients, by volume: 8 parts sand, 4 parts soaked softwood sawdust, 3 parts Type S hydrated lime, and 2 parts Type I Portland cement. The purpose of these tests was to learn the impact of extra sawdust and less sand on the strength of the sample. Cylinder 3 was tested to failure one week later. Cylinder 5 was tested to failure after 30 days.

Here is a summary of the tests. Paul's original Cylinder Compression Test Report is on file at Earthwood Building School.

Cylinder Compression Test Summary

Cylinder Number	Age in Days	Design Mix*	Max. Load in Pounds	Compression Strength, PSI	Type of Fracture
1	7	9-3-3-2	22,000	778	D
6	30	9-3-3-2	35,000	1238	D
2	7	9-3-3-2-w	16,500	584	E
4	30	9-3-3-2-w	24,500	866	E
3	7	8-4-3-2	7,000	248	E, D
5	30	8-4-3-2	23,000	813	D

* The proportions, in order, refer to sand, soaked sawdust, lime, and Portland cement. The "w" refers to samples containing a wood insert at the approximate center of the cylinder mold.

With all three types of sample, the 30-day tests are stronger on compression than the 7-day tests, as would be expected: Cylinder 6 gained 59 percent in strength compared to Cylinder 1. Cylinder 4 was 48 percent stronger than Cylinder 2. And Cylinder 5 was 228 percent stronger than Cylinder 3. It is not expected that much additional strength would be

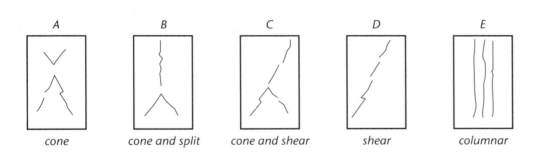

30.1: Typical fractures in test cylinders.

gained by leaving the cylinders to cure for a longer period. The high percentage gain in strength of Cylinder 5 over Cylinder 3 might be explained by the very weak compression strength of Cylinder 3 after 7 days. Samples 3 and 5 contained a relatively high percentage of soaked sawdust in the aggregate (4 parts out of 12, or 33.3 percent of the aggregate) when compared with all the other samples (3 parts out of 12, or 25 percent of the aggregate). The higher sawdust content may be responsible for the lesser compression strength, particularly on the 7-day test (Cylinder 3), where the influence of the sawdust in retarding the set of the mortar is still quite pronounced.

Cylinders 2 and 4 used the same mortar as Cylinders 1 and 6. The difference was that the wooden disk insert, already described, was placed at the center of the sample. The top and bottom mortar cylinders, then, were only about 5¼ inches high, which could explain the columnar failure of these testing cylinders, as opposed to the predominantly shear failure of all the others. Even with the wooden insert, intended to simulate cordwood masonry and the use of wooden plates under rafters, these samples were stronger on compression than Cylinders 3 and 5, which had the higher sawdust content.

The tests seem to support the view that the soaked sawdust admixture accomplishes the intended purpose of retarding the mortar set, thus reducing the incidence of mortar shrinkage cracking. More sawdust retards the set even longer, but at the cost of strength. The 30-day test on Cylinder 6 (9 parts sand, 3 parts sawdust, 3 parts lime, 2 parts Portland cement) is 52 percent stronger than the 30-day test of Cylinder 4 (8 parts sand, 4 parts sawdust, 3 parts lime, 2 parts Portland cement). The 30-day compression strength of all the samples tested is way beyond what is necessary to support even the heaviest cordwood home. The two-story, load-supporting cordwood walls at Earthwood weigh about 2,000 pounds per square foot (or 14 PSI) with a fully saturated earth roof load and a 70-pound (32-kilogram) snow load.

Incidentally, many years ago, when Jaki and I first caught on to the retarding characteristics of soaked sawdust, we tried an aggregate proportion of 10 parts sand, 2 parts sawdust. We wanted to know how little sawdust we could use and still have the desired non-shrink quality. We found that the 10 parts sand and 2 parts sawdust mortar did show shrinkage cracks, so we decided that the 9 parts sand and 3 parts sawdust recipe was optimum. (See also Chapter 11 about the use of commercially available cement retarders.)

Cordwood Masonry in a Seismic Three Zone

Cordwood masonry is strong on compression — way beyond the compression strength required of any bearing wall. But cordwood masonry is not strong on tension. What is the difference?

Compression is the ability of a material or system (in this case, a cordwood wall) to bear vertical loading. Imagine loading a brick until it crushes. Most solid things are fairly strong on compression. Even Dow Styrofoam® Blueboard™ can support 5,600 pounds per square foot (39 psi) with only 10 percent deflection (compression). Tomatoes and your left thumb are not particularly strong on compression, as the impact load from a hammer will readily demonstrate.

Tension is the opposite of compression. It is the ability of a material or system to hold together when it is being pulled apart from opposite sides. Ropes, wire, rebar, and beams are measured in terms of their tensile strength. Masonry is strong on compression but not so strong on tension. Concrete and stone masonry get some tensile strength because of the chemical bond holding the aggregate (or stones) to the cement, but reinforcing bar is placed in concrete to greatly increase its tensile strength. Cordwood is probably the weakest masonry system on tension because there is virtually no chemical bond between the mortar and the log-ends, only a weak friction bond.

A regular load-bearing cordwood wall provides a good "reactionary thrust" to vertical loading because of its compression strength, but during an earthquake, other thrusts are inflicted upon the system. As the building begins to oscillate under the Earth's lateral movements, a sideways thrust is imparted to the wall. First one side and then the other are subjected to tensile stresses, until finally the wall topples over. This is why mud brick buildings perform so poorly in Mexico and other parts of the world during earthquakes, causing great loss of life. On a recent trip to Peru, in an area subject to relatively frequent and strong quakes, I noticed that new mud brick buildings were framed out first in a strong concrete post and beam frame, with plenty of reinforcing bar. These new buildings have a much higher tensile strength. Even if a compartmentalized panel of mud bricks topples over, it is unlikely that serious injury or loss of life will occur.

A similar approach can be taken with cordwood masonry in areas of high seismic risk, except that instead of a concrete post and beam frame (used in parts of Peru because of the lack of trees), a wooden post and beam frame can be employed. The various components of this frame must be tied to each other with either traditional timber framing methods or by the use of metal fasteners (truss plates, floor post brackets, joist hangers, etc.) made for the purpose. Again, in an earthquake, the building will oscillate, but the cordwood panels are small enough that they are unlikely to shake loose of their surrounds. We strongly recommend the inclusion of a wooden key piece attached to the sides of the posts where cordwood is placed.

The key piece is firmly nailed or screwed to the post so that it corresponds to the middle third of the wall. The friction bond of the mortar wrapped around the keyway greatly increases the effective value of the cordwood panel's tensile strength. A determined mule might kick a hole in a cordwood barn wall panel fastened with such a key piece, but without it, he might kick the whole section out in one piece.

30.2: Post with attached key piece.

Richard Kovach and Dawn Danielson built their "Earthwood West" house in Carlsborg, Washington in the late 1980s. Although they had our architect-stamped plans for Earthwood, the Clallam County Building Department insisted upon four changes to accommodate the seismic Zone Three code requirements. Writing in the CoCoCo/94 Collected Papers (*Earthwood Structure in Washington State: Code Issues*, pages 98-103), Kovach and Danielson list the changes. Here is a summary.

1. *Footings.* Twice the rebar and much deeper footings than required in New York State.
2. *Buttresses.* The buttresses at the original Earthwood, designed by an English engineer friend of ours to resist the lateral load of the earth-sheltering, had to be extended to the roof line at Earthwood West and be more heavily reinforced.
3. *Underground block wall.* The below grade block wall had to be "locked" to the footing with L-shaped pieces of #6 rebar and the blocks themselves locked to each other with vertical rebar and frequent use of "bond beam" courses of block. A "knockout bond beam block" allows the placement of horizontal rebar slushed with concrete, greatly increasing resistance to lateral loads such as earth pressures and earthquake oscillation.
4. *Load-bearing external wall (cordwood).* As this is the part that will be applicable to any normal above grade cordwood home in a seismic area, I'll let Richard and Dawn tell the story:

"Code requires that all load-bearing masonry be reinforced. Since we could not envision threading rebar into our cordwood masonry matrix (at least not without losing our sanity), we chose to use posts and beams for the external wall loading. Douglas fir 8-by-8's were used for all of the post and beams. Rafters were 4-by-10's, and joists were 4-by-8s, also Douglas fir. All framing wood was purchased from a local mill, selected from 'butt ends' near the ground for clarity and minimal knots, and had to be graded by a certified inspector. All pieces were found to be #1 grade or better. The cost of the timbers was $2700 and the inspection was $35

"First-story posts were connected to the footing by bolting them into Simpson CB88 connectors, which were placed before pouring the footing. Each post was bolted to its nearest beam with two, ½-inch-by-8-inch lag bolts at a 45° angle. Adjacent beams were connected together with truss plates. Care was taken to ensure plumb and level on all posts and beams, and our efforts paid off. For a couple of first-timers who had built nothing more complicated than model airplanes, we had no unpleasant surprises."

Even in a non-seismic area, the use of a post and beam framing for load support has a lot going for it. Besides enabling you to get the roof on early — and doing the cordwood work under cover — code enforcement officers love it.

The Thermal Performance of a Cordwood Wall

As Kris Dick pointed out in Chapter 27, a cordwood wall is more than just insulation. The thermal mass characteristics of this building system are equally important. Earthwood stays a steady temperature, summer and winter, and I credit the thermal mass for this. And I can think of no other building system that so effectively combines insulation with thermal mass. Log-ends have characteristics of both. But it is the mortared portion of the wall, with its placement of thermal mass on each side of the insulation that really shines in this regard. The inner mortar joint stores warmth in the winter and coolth in the summer. The outer mortar joint acts as a giant heat capacitor as well, protecting the interior from rapid temperature fluctuations.

A code official may insist upon a certain R-value in the wall, based upon energy codes that may be in effect in your area. In New York, R-19 is required for walls. By using white cedar, the highest rated of the common North American woods for insulation value at R-1.5/inch on side-grain (probably closer to R-1 on end-grain), we were able to achieve an insulation value of about R-19.6 for our 16-inch (40-centimeter) wall. But how do we arrive at this figure?

We measured the cross-sectional area of each and every log-end in a typical panel at Earthwood, a six-square-foot panel of cordwood in son Darin's bedroom. Almost exactly 3.6 square feet of the panel consisted of log-ends, leaving, by subtraction, 2.4 square feet of mortar. So the build quality that Jaki and I employ yields a wall that is 60 percent wood and 40 percent mortar matrix, in cross-section. With cedar on end-grain, we are only getting about R-16 for our 16-inch thick walls. But the mortared portion of the wall does a lot better. The 6 inches (15 centimeters) of light sawdust between the mortar joints at R-3.5/inch is worth about R-21. But we also gain at least a nominal R-2 per sawdust mortar joint, taking into consideration the two extra heat transfers from one type of material to another. The mortared portion of the wall, then, brings the average R-value of the wall up to R-19.6.

In 1974, a 24-inch-thick (60-centimeter-thick) stackwall structure with an insulated center portion was built at the University of Manitoba by Drs. Allen Lansdown and Arthur Sparling. The walls were rated at R-24, according to author David Square, who interviewed the builders for his article, titled "Poor Man's Architecture" (*Harrowsmith Magazine*, #15, May 1978).

Dr. Kris Dick, P.E., in Chapter 27, assigns a value of R-20 for 24-inch-thick poplar walls but says, "In many cases, however, the actual R-value seems to exceed R-20 by a large margin," and "Experiments with full-scale stackwall buildings have indicated effective R-values considerably higher than calculated R-values."

Finally, if you want a cordwood home with incredibly high R-values of around R40, and with a vapor barrier, use Cliff Shockey's Double Wall Technique of Chapter 4. His system, too, places excellent thermal mass on each side of the insulation space.

In summary, cordwood houses can easily meet R-value codes, and in so doing yield thermal characteristics not present in a typical low mass insulated wall — benefits like heat storage and a steady temperature curve. If the fires go out in a cordwood home, it takes quite a while to notice it. In a home with insulation only and few BTUs in reserve, temperatures will drop quite quickly.

Although this is the last formal chapter, it is not the end of the book. What follows has some good information, so stick around a while longer.

Afterword
Where We Go From Here

Rob Roy

I HOPE YOU HAVE ENJOYED THIS TRIP through the world of cordwood masonry. Indeed, various authors have given us a good look at the state of the art. Now what?

Just as watching an exercise video will not take inches off your belly, reading this book won't get a cordwood wall built. The book can only guide you. You have to do the work.

Some of the authors started with a practice building — a good idea. Jaki and I advise our students to build something small, like a well-house, garden shed, sauna, or even a temporary shelter (as discussed in Chapter 26). If you can't build the practice building — or don't enjoy it — well, you've learned something valuable: cordwood is not for you. But grandmothers, children, and beavers all build cordwood buildings successfully and enjoy it, too, although some of the beavers with whom I'm acquainted tend toward being workaholics.

After gaining practice, skills, and a sense of how long it takes you to build (everyone is different in this regard), you are far better placed to design and build your home. You will be far more likely to build something manageable. If you are unsure about structure, follow the advice of Cordwood Jack Henstridge. Build a model. You may recall that Larry Schuth, way back in Chapter 17, discovered that trying to build his balsa model revealed some structural problems, which he was able to work out with an architect.

Give yourself plenty of time for the permitting process. Stay on good terms with the code enforcement officers. Keep a sense of humor.

Where do Jaki and I go from here? Well, we are already talking about building a creative little guesthouse on a piece of property near Earthwood. We'll do it as a workshop project, but we are also interested in making it a showcase for the art of cordwood masonry. We may need to take Val Davidson on as a consultant. We'll probably keep building and teaching about cordwood until we can't lift a five-pound log-end. Old cordwood masons never die — although they may get laid up.

And cordwood masonry in general? The future has never been brighter. In 2004 or 2005, there will be another Continental Cordwood Conference, probably in the Midwest. The exact location had not been chosen at press time. You can get the latest news on this by visiting either the Earthwood or the Daycreek websites (see the Bibliography).

The really exciting news is that on June 20, 2002 near Stevens Point, Wisconsin, a number of cordwood builders and writers, including several authors from this book, met together and started the Cordwood Builders Association. I wish we'd done this ten years earlier, but better late than never. Our goals are to gather together all the information available on cordwood masonry, find out what we've got, and find out what we need. Not only do we want to provide a place for getting technical help for builders, but we also want to provide the kind of authoritative information most often requested by code enforcement officers. With help from major universities in both Canada and the United States, we hope that we will soon have approved tests on such matters as load-bearing characteristics and R-values for cordwood masonry. Of course, we already know that cordwood performs very well in these regards, but some building inspectors really need to see those test results. This Cordwood Builders Association is only a few months old as we go to press, so again, please visit the Earthwood or Daycreek websites for the latest contact information. Or write Earthwood Building School, 366 Murtagh Hill Road, West Chazy, NY 12992.

Happy stacking!
Rob Roy
West Chazy, New York

Bibliography

ANNOTATED CORDWOOD MASONRY BIBLIOGRAPHY

(Editor's Note: I have confined this list to materials still current at press time. Out of print cordwood books, including my own, have been replaced by better and more up-to-date works. If you are on low budget, make use of your local library. If they haven't got a book you want, suggest that they purchase it, or ask them to get it for you on inter-library loan.)

CORDWOOD BOOKS

Dick, Kris and Allen Landsdown. *Stackwall: How to Build It*, 2nd ed. A and K Technical Services, 1995. PO Box 22, Anola, Manitoba, Canada ROE OAO. This one revises the original 1977 edition and has lots of new information. Accents stackwall corners and the gravel berm foundation. Well-illustrated, 114 pages, large paperback, spiral bound. Co-author Kris Dick also wrote Chapter 26, herein.

Flatau, Richard. *Cordwood Construction: A Log End View*. Self-published, 2002. This one recounts the author's adventures in building a 3-bedroom, 2,100-square-foot (195-square-meter) mortgage-free cordwood home within a post and beam frame. Lots of "frequently asked question" answered. Well illustrated. Revised and updated in 2002. $14 postpaid. Richard Flatau, W4837 Schulz Spur Drive, Merrill, WI 54452; Phone: (715) 536•3195; E-mail: flato@aol.com.

Henstridge, Jack. *ABC: About Building Cordwood*. Self-published, 1997. This is "A 25-year collection of papers, presentations, notes, and what-not." And it's written in Jack's inimitable homespun and entertaining style. 72-page large paperback. Available from Jack Henstridge, 1149 Route 102, Upper Gagetown, New Brunswick, Canada E5M 1T3; Phone: (506) 488•2477.

Roy, Rob. *Complete Book of Cordwood Masonry House-building: The Earthwood Method*. Sterling, 1992. The first half of the book is a basic cordwood primer, covering all three styles: post and beam, stackwall corners, and curved wall. The second half describes the construction of the Earthwood home, from soup to nuts. Large format, 256-page book, fully illustrated.

Shockey, Cliff. *Stackwall Construction: Double Wall Technique*. Self-published, 1993. Expanded to 96 pages and revised in 1999, the book focuses on Cliff's award-winning, energy-efficient double wall design, which he describes in Chapter 5 herein. The new material covers his sauna, in-floor heating, his 1997 addition, and more. Cliff Shockey, Box 193, Vanscoy, Saskatchewan, Canada SOL 3J0; Phone: (306) 668•2141.

Books with One or More Chapters about Cordwood Masonry

Chiras, Dan. *The Natural House: A Complete Guide to Healthy, Energy-Efficient Environmental Homes*. Chelsea Green, 2000. The subtitle says it all. The chapters on cob and cordwood masonry are both very good, like the rest of the book.

Kennedy, Joseph. F., Michael G. Smith, and Catherine Wanek, eds. *The Art of Natural Building*. New Society Publishers, 2001. PO Box 189, Gabriola Island, British Columbia, Canada, V0R 1X0. A well-detailed overview of a variety of natural building methods. Rob Roy wrote the section on cordwood masonry. 288 pages.

McClintock, Mike. *Alternative Housebuilding*. Sterling, 1989. Although a little outdated now because of all the new work done in the alternative building field in the '90s, there is still a lot of good information here. The chapter on cordwood masonry is well researched and quite interesting.

Roy, Rob. *Mortgage Free! Radical Strategies for Home Ownership*. Chelsea Green, 1998. This one expands on Chapters 2 and 26 herein and also includes a cordwood case study in Washington State.

—————. *The Sauna*. Chelsea Green, 1996. A cordwood masonry sauna is a great starter project and delivers a genuine Finnish sauna experience. This book is about sauna, but three of its seven chapters are really about cordwood. Chapter 3 is about building a post and beam "Log-End Sauna," and Chapter 4 is about building a round cordwood sauna. Other chapters deal with sauna lore, stoves, and how to take a sauna. Fully illustrated, 198 pages.

Wrench, Tony. *Building a Low Impact Roundhouse*. Permanent Publications, 2001. Distributed in the US by Chelsea Green. This book is a greatly expanded version of Chapter 19, well-illustrated, and spiced with the author's philosophy of living gently with nature.

Cordwood Videos

Basic Cordwood Masonry Techniques. Rob and Jaki Roy. Chevalier/Thurling Productions and Earthwood, 1995. This 88-minute video is like a cordwood workshop in a can and covers barking the wood, estimating materials, mixing mortar, building the walls, pointing, laying up window frames, stackwall corners, and more.

Cordwood Homes, Presented by Rob Roy. Rob Roy. Chevalier/Thurling Productions and Earthwood, 2000. A 75-minute tour of a variety of cordwood homes, inside and out. Includes interviews with several authors from this book, including Jack Henstridge, Kris Dick, Geoff Huggins, Scott Carrino, and Jim Juzcak.

Magazines with Frequent Articles on Cordwood Masonry

BackHome Magazine. PO Box 70, Hendersonville, NC 28793; Phone: (828) 696•3838. Over the years, *BackHome* has published quite a number of articles about cordwood masonry homes and the use of cordwood masonry with sauna construction. Publisher Richard Freudenberger moderated the 1994 and 1999 Continental Cordwood Conferences. Back issues are available at reasonable prices. Call to find out which ones have cordwood articles. See also the Daycreek website, following.

Mother Earth News. 1503 SW 42nd Street, Topeka, KS 66609; Phone: (785) 274•4300. The earliest cordwood articles in a major magazine appeared in *Mother* and helped the cordwood renaissance get a jumpstart back in the '70s. They have had several cordwood case study articles over the years. Back issues are available but expensive. The Daycreek website, listed below, has most of the old Mother Earth News cordwood articles online.

Cordwood Masonry Websites

www.cordwoodmasonry.com
This is the website for Earthwood Building School, which specializes in cordwood masonry instruction. Many of the books and videos listed above are available through this site.

www.daycreek.com
Cordwood builder Alan Stankevitz maintains the Daycreek website, which focuses on cordwood masonry in general, with lots of good articles, including articles from back issues of *Mother Earth News* and *BackHome Magazine*. There is a detailed diary of his own two-story, round house project, which makes use of

paper-enhanced mortar. This is also the best cordwood masonry chat room on the Web, with many of the most experienced cordwood builders contributing. This is the place to keep up with "The State of the Art."

www.groups.yahoo.com/group/AllCordwood
Another chat room on cordwood masonry. Worth a look.

www.greenhomebuilding.com
Green Homebuilding is a fine website by Kelly Hart and covers all sorts of natural and vernacular building, including a section on cordwood masonry. The "Ask Our Experts" feature enables the visitor to ask questions about several different natural building techniques. Rob Roy fields the cordwood questions.

www.newsociety.com
This is the site for New Society Publishers. If you have enjoyed this book, you'll probably be interested in several of their other titles on building with nature.

Cordwood Instruction

Earthwood Building School. Since 1979, Earthwood (366 Murtagh Hill Road, West Chazy, NY 12992) has been conducting cordwood workshops in northern New York and around the world. Earthwood also serves as a clearinghouse for all things cordwood, including most of the books in this Bibliography. Write, or call (518) 493•7744 for the latest books and videos list and workshop brochure. Or visit the Earthwood website at: www.cordwoodmasonry.com.

Glossary of Terms

Air infiltration – The transfer of air though the fabric of a building. Log-ends that shrink a lot are sources of air infiltration and thus promote heat loss by convection. A polythene vapor barrier greatly reduces air infiltration.

Bed – In masonry, the mortar upon which a brick, block, stone, or log-end is laid.

Bottle-end, bottle-log – A mixture of two bottle or jars joined together to make a glass masonry unit for the admission of light. (See Chapter 8.)

Built-up corners – In cordwood masonry, a corner system by which corners are constructed of regular wooden blocks, called "quoins," laid up in an alternating crisscross fashion. Also known as "stackwall corners." See also "Lomax Corners."

Cement – The hardening and strengthening agent in mortar and concrete. See also "Portland cement" and "masonry cement."

Cement retarder – One of a number of commercially available products used as additives to concrete or mortar for the purpose of slowing the set of the material.

Checking – The natural splitting of a log-end (or any piece of wood), resulting from rapid drying. A presplit log-end often has hairline cracks, without a primary check that goes all the way through from end to end. A single large check is a common condition with cylindrical log-ends.

Cob – A mixture of sand, clay, straw, and water, used to build walls. Can also be combined with log-ends to build a cordwood wall. See next entry.

Cobwood – A new term coined by cordwood and cob builders, referring to a cordwood masonry wall tied together with cob instead of mortar.

Concrete – A mixture of sand, stone aggregate, Portland cement, and water. When concrete sets, it makes a strong wall, slab, deck, or foundation material. Not to be confused with "mortar."

Cord – A unit of measure for stacking and purchasing firewood or pulpwood. While, technically, a cord of wood should refer to a "true," "full," or "real" cord of 128 cubic feet (3.6 cubic meters), the term now commonly refers to any stacked pile of wood with a sectional area of 32 square feet (3 square meters), normally four feet high and eight feet long. If the stack is also four feet wide, it will be a true cord of 128 cubic feet. See also "face cord."

Cordwood masonry – A wall-building system in which short logs, often called "log-ends," are laid up transversely in the wall within a special mortar matrix, much as a cord of firewood is stacked. Also, "stovewood masonry," "stackwall," "firewood wall," and the like.

Double Wall Technique – A thermally efficient wall system, made from two separate cordwood masonry walls separated by a fully insulated cavity. Although at least one double wall barn from the 1930s has been identified in Michigan's Upper Peninsula, double wall for housing was developed by Cliff Shockey in 1977. (See Chapter 4.)

Drawknife – A sharp, single-edged metal blade with a handle at each end of the cutting edge. Used mainly for shaping wood, a drawknife can also make a good barking tool when all else fails. (See Chapter 3.)

Face cord – A stack (also "rank," "rick," or "run") of wood four feet high, eight feet long and a certain agreed-upon thickness: 12 inches or 16 inches (30 centimeters or 40 centimeters), for example. The face cord is a convenient measure to use in determining material requirements for a cordwood project. It is important that the buyer knows exactly what the seller means by the term "cord."

Firewood walls – An archaic term for cordwood masonry walls.

Floating ring beam – A ring of concrete footings floating on a pad of percolating material. See next entry.

Floating slab – A foundation method, whereby a concrete slab is "floated" on a pad built up from runs of good percolating material such as coarse sand, gravel, or crushed stone. A favorite of Frank Lloyd Wright, the floating slab is an economic choice for a cordwood foundation in areas of deep frost. Not recommended on expansive clay soils.

Footer, Footing – Foundation base for a wall or building.

Girder – A major horizontal beam, which supports floor joists or roof rafters.

Lintels – Wooden timbers that carry wall load over doors or windows.

Lime – In masonry, a white caustic powder added to mortar to improve its plasticity. "Mason's lime" (also called "builder's lime," "hydrated lime," or "Type S lime"), is made by converting limestone by heat. "Agricultural lime," which is non hydrated, is used as a soil conditioner in agriculture and is not suitable as a mortar additive.

Log-ends – The individual short logs, butts, blocks, ends, or pieces of wood used as masonry units in a cordwood wall. Log-ends are most commonly used transversely in the wall, where, with their end-grain exposed, they "breathe" very well, greatly reducing the danger of wood deterioration through rot.

Lomax Corners – Built-up corners made from pre-built corner units, instead of individual quoins. Named for Gary Lomax. (See Chapter 5 for details.)

Masonry cement – A cement and lime mixture that has become popular with modern masons. There are several types, with varying characteristics.

Mortar – A mixture of sand, cement, and water used for laying up masonry units such as bricks, blocks, stones, or log-ends. Sometimes other ingredients are added for certain purposes. Colloquially know as "mud."

Mortar mix – Common term for a dry, bagged, premixed mortar product, usually about three parts sand to one part masonry cement. Just add water for a good brick or block mortar. Not to be confused with bags of masonry cement, which contain no sand.

Mud – Slang for "mortar."

Panel – In cordwood building, a section of masonry enclosed by posts and beams.

Paper-enhanced mortar (PEM) – A mortar with a high recycled-paper content. (See Chapters 14 and 15.)

Papercrete – A material made from paper, cement, and water, used for building. The density and strength of papercrete varies widely with the recipe and whether or not sand or other admixtures are included. See also "paper-enhanced mortar (PEM)."

Peeling spud – A chisel-like tool made for removing bark. Many cordwood builders have made successful spuds by mounting a wooden handle to the leaf spring from an old truck. A heavy pointed mason's trowel makes a pretty good peeling spud, too.

Plate – In cordwood masonry, wooden planking (typically two inches thick) used to distribute joist or rafter load onto the cordwood wall. The plate greatly distributes the concentrated load of these members onto the cordwood wall. The plate can also tie corners together and provide a surface upon which to fasten floor joists or rafters.

Plate beam – In a post and beam frame, the topmost horizontal member; the top of a cordwood masonry panel. See also "plate."

Pointing – The process of smoothening the mortar between masonry units. Also called "tuck-pointing" or "grouting." With brick or block work, the term "raking" is also used.

Pointing knife – A tool used for pointing. Can be made by bending the last inch of a smooth kitchen butter knife to an angle about 15 degrees raised from the plane of the knife blade.

Portland cement – A strong, unmixed cement used in concrete, made by burning a mixture of limestone and clay or other materials. Type I is the basic type, with standardized strength characteristics.

Proud – In masonry, the opposite of recessed. Protrusive, as in, The log-ends sit proud of the mortar background.

Quoins – In cordwood masonry, the individual blocks of wood used in the construction of built-up corners, usually made from regular dimensional material, such as 6-by-6-inch timbers. In stone masonry, squared stones used in corner construction.

R-value – A measure of insulation value in building materials. The higher the R-value number, the greater the insulation. Materials are often measured in terms of R-value per inch. Extruded polystyrene, for example, is about R-5/inch.

Rank, Rick, Run – See "face cord."

Random rubble pattern – In cordwood masonry, the use of a variety of sizes and shapes of log-ends, distributed randomly in the wall.

Retarder – See "cement retarder."

Ridgepole, Ridge beam – The major carrying beam or girder of a roof system, supported by posts.

Sill plate – A wooden plate, often pressure treated, which caps the top of concrete footings, a poured concrete wall, or a block wall. Also "toe-plate" or "sill."

Sills – Heavy horizontal wooden timbers sometimes installed beneath window framing. See also "sill plate."

Stackwall – Cordwood masonry, particularly in Canada.

Stackwall corners – Same as "built-up corners."

Stovewood masonry – Same as "cordwood masonry." The term is most commonly encountered in historical articles and is seldom, if ever, used in reference to cordwood structures built since 1960.

Thermal mass – The capacity of a material to store heat. Generally, a material's thermal mass characteristics are inversely proportional to its insulation characteristics.

Toe-plate – See "sill plate."

Index

A

Abel, Mike (Missouri), 82
Acryl-60, mortar bonding agent, 53
Add-ons, strategy for, 191–192
Adkisson, George (Texas), 127–132
Alaskan Small Log Mill, chainsaw attachments, 59
Appalachia-Science in the Public Interest (Kentucky), 99

B

Bailey's Woodsman's Catalog, chainsaw equipment, 59
Bark, removing. *see* Debarking wood
Basement, cost and use of, 15, 43, 190
Benchsaw. *see* Saws
Bottle-logs
    assembly, 65–69
    choice of color and shape, 63
    placement in wall, color, light and shadows, 64–69, 73
Building codes and permits, 108, 123–124
    approval criteria, 199–201, 207
    building to code, 197–199, 203–212
    designing for, 17, 203–204, 208–209
    NYS, 203–205
    post and beam framework, 204
    structural integrity, 127, 194, 209
Building inspectors, working with, 193–201, 207, 211
Building sites. *see* Site selection
Buildings, shape of. *see* Round buildings
Built-up corners. *see* Lomax corner

C

Canadian Inventory of Historic Buildings in Quebec, 7
Carlson, John and Carrino, Scott, 207–212
    Community Round House (New York), 167–176
Caulking. *see* Gaps
Cement. *see* Mortar
Cement retarder, 28, 85, 90–91, 149, 179
    testing, 89–90
Chainsaw. *see* Saws
Chainsaw mill, Bailey's Woodsman's Catalog, 59
Cistern, 100
Clay. *see also* Cob construction
    shake test, 144
    substitute for mortar, 134, 139–145
Cob construction
    cordwood and, 133–138
    North American School of Natural Building (Oregon), 139–140
Cobwood construction
    clay mortar, 134, 139–145
    clay shake test, 144
    cobwood vs solid cob, 144
Code, building. *see* Building codes and permits
Community Round House, 167–176

Concrete blocks, below grade foundation, 17
Concrete slab
    floating, 38, 127–128
    planning for utilities, 79
    slowing cure, 85
    under-floor radiant heating, 38–39
Condensation. *see* Moisture management
Continental Cordwood Conference (CoCoCo), 222
Cordwood and cob, 133–145
Cordwood and stone, 158–160, 177
Cordwood Builders Association, 222
Cordwood wall construction. *see also* Post and beam framework
    added to mobile home (Kentucky), 99–101
    building process, 31–34, 124, 134-5. *see also* Corners; Fraser Frame
    compression, 216
    designing log pattern around doors and windows, 72–73
    double wall, 39–42
    durability, 195
    energy efficiency, 5, 19, 195, 210–211
    environmentally friendly, 101, 133, 136, 210
    family benefits of, 125, 130–131, 176
    framing techniques, 5, 50–57, 124, 136–139, 174
    history in North America, 4, 6, 195
    load bearing walls, 54, 56, 169, 208, 216–218
    mortgage free, 164–165, 183–192
    steel framework, 169, 174
    straightening walls, 172–173
    straw bales and, 135
    strength, 213–218
    thermal mass, 218–219
    wall thickness, 26
Cordwood with steel framework, 169–174
Corners
    angles, framing, 58–61
    Lomax corner, 44–47
    round posts, framing with, 50–59
    stackwall, 5, 61
Cutting table, 27, 74–77, 123

D

Daratard 17, 89
Davidson, Jim and Valerie
    Marshwood home (B.C.), 63–69, 71–74
Davidson, Valerie, bottle designs, 63–69
Debarking wood, best time for, 23–25, 161, 170
Decay of wood
    prevention of, 22–23, 129, 194, 197
    wood preservative, 101, 155
Degree days, calculating fuel requirements, 26
Dick, Dr. K.J. and Lansdown, Professor A.M., 193–201
Doors, 17, 100
Double wall. *see also* Cordwood wall construction; Mortar
    insulation value, 42
    sheathing, use of, 40
    technique, 39–42
    vapor barrier, use of, 41
Drainage around building. *see also* Foundation, 17
Drawknife, 23–24, 161

E

Earth on roof. *see* Roofing
Earthquake, limiting damage, 216–218
Earthwood Building School
    building cost, 19
    construction of, 18
Earthwood House. *see also* Log End Cave; Log End Cottage; Mushwood House (B.C.)
Electrical installation
    flexible conduit through walls, 52,

80–83, 199
incoming power hookup, 204
National Electrical Code, 79, 204
Wiremold (metal conduit tubing), 81
Energy conservation. *see also* New York State Energy Conservation Code
   energy efficiency, 5, 19, 101, 195
   shutters, insulated, 164
   thermal mass and, 195, 218
Evans, Lanto and Smiley, Linda (Oregon)
   North American School of Natural Building, 139–140

F
Facade, esthetically pleasing, 39
Family benefits of cordwood housing, 125, 130–131, 176
Filling gaps. *see* Caulking
Fire
   escape route, 17
   protection from, 197–199
Fireplaces
   central concrete, 104
   wood burning, 129
Floor layout, 57, 188–189
   thickness of walls, 15, 26, 119. *see also* Polygons; Round buildings
Floors. *see also* Heating and cooling system; Insulation
   beaten earth, 138
   wood, 119, 135
Floors, multiple, 50–51, 127, 175
Formless Concrete Construction Company, 9
Foundation. *see also* Planning
   building on bedrock, 124
   Community Round House, 167–176
   concrete blocks, 43, 100
   concrete cracking, 194
   cordwood above stone masonry, 177
   floating concrete slab, 38, 127–128
   footings, poured, 119, 217
   in-floor hot water heat, 38
   limiting earthquake damage, 216–218
   limiting hurricane damage, 130
   load bearing, 119, 127, 194
   ring beam over crushed stone, 104
   rubble trench, 159
   slip-form method, use of, 158–159
   structural strength of, 194
Framing. *see also* Polygons; Post and beam framework
   radial floor joist system for second story, 51
   steel framework, 169, 174, 209
   temporary diagonals, 49–56
   temporary post, 52
   windows, 100, 127, 136–139
Fraser, Bunny and David (Bear)
   Hutchnden House (Ontario), 49–56
Fraser frame (external frame)
   infilling post and beam frame, 50–57
Fritsch, Al and Kieffer, Jack
   mobile home surround (Kentucky), 99–101
Fungi (wood rot), prevention of, 22–23, 129

G
Gaps, filling. *see also* Mortar
   caulking, timing of, 195
   latex and silicone caulking, 94, 96
   other caulking materials, 94–95, 148, 161
   refinishing with mortar slurry, 96–97
Government inspections. *see also* Building codes and permits, 211

H
Hardwoods. *see* Wood
Heating and cooling system. *see also* Insulation; Solar design; Stove; Windows
   concrete heat sink, 104
   hot water heating, 38–39

in-floor radiant heating, 38
thermal mass, heat storage in, 218–219
ventilation, 163
window shutters, insulated, 164
woodstove, 19, 134, 163, 205
Hebel, Hans (Chile), 147–150
Henstridge, Jack (New Brunswick), 43–47
Higgins, Wayne and Marlys
Stonewood House (Michigan), 117–120
History in Europe, 8–9, 195
History in North America, 4, 6, 195
Huber, Tom, stone masonry and (Michigan), 157–160
Huggins, Geoff & Louisa (Virginia), 93–97
Hurricanes, limiting damage, 130

I

Insects and pests, prevention of damage, 31, 128, 130
Inspections by code enforcement officers. *see also* Building codes and permits, 199–201, 203–205, 207, 211
Insulation. *see also* Heating and cooling system
caulking, 195
energy efficiency, 195
fiberglass, 30, 39
footing, 17
optimum R-values, 195, 218–219
R-value of insulation, 31–32, 195
R-value of log-ends, 26, 195
sawdust, 30–31, 179
sheathing, value of, 40
shredded beadboard, 31
Styrofoam board, 100
vapor barrier and, 40–41, 196–197
window shutters, insulated, 164

J

John Mecikalski General Store, Jennings, Wisconsin, 10

Juczak, James S. and Krista
18-sided home (New York), 103–108

K

Ketter-McDiarmid, Stephen and Christine, 177–180
Kieffer, Jack (Kentucky), 99–101
Kwiatkowski, Thomas M. (New York), 203–205

L

Land selection. *see* Site selection
Las Ventanas Naturales (windows), 147–148
Light fixtures. *see* Electrical installation
Lime. *see* Mortar
Lind, Olle (Sweden), 151–154
Log butt. *see* Cordwood wall construction
Log End Cave, design, 16
Log End Cottage, design, 14–15
Log-ends. *see also* Bottle-logs; Wood, 43
building process, 31–34, 124–126. *see also* Corners; Fraser frame
building with cob, 134, 139–145
criteria for log-ends, 21–23, 128, 133, 160, 170
cutting into log-ends, 27, 74, 123. *see also* Saws
debarking or skinning, best time for, 23–25, 129, 161, 170
drying time, 22–23, 85, 128, 154–155
estimating quantity, 24–25
keeping dry, 22–23, 129, 160, 173–174
masonry sealers, 41, 94–95
moisture absorption, 85–86, 154, 156
pointing logs, pointing knife, 32–33, 125, 161
R-value of log-ends, 26, 195, 218–219
shrinkage/expansion, 21–23, 129, 154–155
size of log-ends, 27, 31, 71
splitting to reduce checking and shrinkage, 27

wood preservative and, 101, 155
Lomax corner, 44–47, 61
Lomax, Gary (New Brunswick), 44
London, Bev. (New Brunswick), 43

M

Marshwood, 64–65
Masonry sealers, 94–95
Masonry under cover, 15, 177–178
Mikalauskas, Paul (New Hampshire), 79–83
Moisture management, 171–172, 174
    limiting exposure to rain, 27, 129
    moisture absorption of wood, 85–86, 154–156
    vapor barrier and condensation, 40–41, 196–197
Moller, Steen, cobwood house (Denmark), 144–145
Mortar. *see also* Pointing
    bonding agent, Acryl-60, 53
    building process, 31–32
    building with cob (clay and straw), 134, 139–145
    cement, 29, 86
    cement retarder, using, 28, 85, 89–91, 149, 179
    "cheaters" for spacing log-ends, 175
    cob (clay) as substitute for, 134, 139–145
    color variation, 17, 161, 179
    compression strength, 209
    compression tests, 213–216
    cracking or fracturing, 153–154, 172, 175
    creating patterns, 175. *see also* Bottle-logs
    drying time, 28, 85–86
    lime and, 29
    masonry cement, 29
    masonry sealers, 41, 94–95, 161
    mixes, 28–29, 40, 119, 128, 152–153, 161, 172
    mixing
        adding water to dry mix, 30
        cement retarder, using, 91
        clay and straw cob, 140–142
        mechanical mixer, using, 55, 171, 179
    paper-enhanced mortar (PEM), use of, 107–113, 161
        paper sludge, 107
        shredded newspaper, 109–111
        shredded white office paper, 112
    pointing
        beautifying wall, 33
        creating bond with mortar, 33
        smoothing mortar, 161
    quality control, 88, 175
    rubber gloves, need for, 30
    sand, quality test, 175
    sand, suitable, 28
    sawdust, types of, 28, 55, 86, 88
    shrinkage
        reasons for, 93
        test of, 87, 93, 213–216
    structural integrity, 209
    tensile strength, 216
Mortgage free. *see also* Planning, 130, 164–165, 183–192
Multiple floors, 50–51, 127, 175
Multi-sided buildings. *see* Polygons; Round buildings
Mushwood House (B.C.)
    design of, 19
    domed ceiling, 20

N

National Electrical Code, 204
New York State Energy Conservation Code
    energy code compliance and exemption, 210
    energy performance, rating, 210
Norris Miller House (Iowa), 6
North American School of Natural Building, 139–140

O

Octagonal framework. *see* Polygons
Oregon cob. *see* Cob construction
Origin of cordwood building, 3–10

P

Paper-enhanced mortar (PEM). *see also* Mortar
  value of, 107, 109–113
Peeling spud, debarking wood, 23–24, 161, 170
Perma Chink, masonry sealer, 95, 161
Permits, building. *see* Building codes and permits
Planning. *see also* Building codes and permits; Foundation; Post and beam framework; Safety; Solar design
  architect's plans, preliminary, 190, 204, 208–209
  building orientation to the sun, 16
  design process, 133–139, 168–170, 221
  electrical installation
    flexible conduit through walls, 52–53, 199, 211
    incoming power hookup, 79–83, 204
    National Electrical Code, 79, 204
    Wiremold (metal conduit tubing), 81
  estimating quantity of log-ends, 24–25
  fire protection, 196–198
  floor area, 15–17, 57, 188–189
  keep it simple, 119, 127, 187–192
  limiting earthquake damage, 208–209, 216–218
  limiting hurricane damage, 130
  masonry under cover, 15, 50, 125, 174, 177–178
  moisture management, 174, 196–197
  mortgage free, 130, 164–165, 183–192
  roof overhang, value of, 125
  shape of building, 187–190
  start small, 165, 186
  structural safety, 127, 194
  wall thickness, 15, 26, 119
Pointing
  beautifying wall, 33
  creating bond with mortar, 33
  smoothing mortar, 161
Pointing knives, 33, 125
Polygons, framing, 49–56, 103–108
  corner angles, 57–61
Portland mix. *see* Mortar
Post and beam framework. *see also* Cordwood wall construction; Mortar
  advantage, 217
  cordwood infilling, 50
  corners and, 44–47, 50–59, 61
  designing pattern around doors and windows, 71–73
  double wall, 39
  limiting earthquake damage, 208–209, 217–218
  limiting hurricane damage, 130
  octagonal framework. *see* Polygons
  raising posts and joists for second floor, 105, 174, 178
  round posts, framing with, 59–61
  structural considerations, 127, 169–170, 194

Q

Quoins, use of, 45–47, 61

R

Rain, limiting exposure of wood, 27, 129, 172
Recycling, 105, 106–112, 160, 191 *see also* Wood; Salvage
Roofing
  construction of, 54–56, 173–175, 177–178
  Earthwood, 18, 56

framing rafters, trusses, 105, 119, 134, 178
roof on before infilling walls, 15, 161, 174, 177–178
roof overhang, value of, 125
shingle, sod or turf, 55–56
steel sheets, 99, 174
temporary, 173
waterproofing membrane, 54, 173
white metal roof, 158

Rot, wood, prevention of, 22–23, 129, 194, 197

Round buildings. *see also* Polygons
building shapes, 187–190
Community Round House (New York), 167
construction of, 133–138, 167–176
curved walls, 6
economy of scale, 190
framing corner angles, 58–61
framing corner angles with round posts, 50–59
framing for 16 sided home, 49, 54
framing for 18 sided home, 54, 103–108

Roy, Rob, 13–20, 21–34, 49–56, 57–61, 71–77, 85–91, 139–145, 183–193, 213–222

Roy, Rob and Jaki (N.Y.). *see* Earthwood House, Earthwood Building School, Log End Cave, Log End Cottage, and Mushwood

S

Safety. *see also* Building codes and permits
fire escape, 198–199
fire, preventing spread of, 197–198
safe construction site, 211
saws, using, 27, 76–77, 161

Salvage, wood. *see also* Recycling, 105, 178
Sand in mortar, quality test, 175
Sawdust for insulation, 30–31

Sawdust for mortar
cement retarder alternative, 91
quality control, 88
shrinkage test, 87
suitable, 28, 55, 86–88

Saws
attachments, 59, 77
cutting log ends, 27, 133, 161
cutting table, 74–77, 123

Scandinavian cordwood buildings, 8–9, 151–155

Schuth, Larry, (Woodland Treat, New York), 121–126

Scraper, debarking. *see* Wood

Sheathing, 40

Shockey, Cliff
double wall technique (Saskatchewan), 37–42

Shrinkage and expansion of log-ends, 21, 27, 129, 154–155

Site preparation, 55

Site selection. *see also* Solar design
features, 18–19, 121, 127, 184–185, 199
southern exposure, 162
wind direction, affect on site, 162

Skedd, Charles
stackwall pumphouse (North Carolina), 155

Skylights, 17
Sod roof. *see* Roofing
Softwoods. *see* Wood

Solar design
site selection, 16, 38
solar energy, 38, 162
thermo mass, heat storage, 163, 195, 218–219
thermo-shutters, 164
white metal roof, 158

Square corners
adaptation for polygons, 57–61
Lomax corner, 43

Stackwall construction. *see* Cordwood wall construction
Stairs, 119
Stankevitz, Alan, (16-sided house, Minnesota), 109–113
Stone and cordwood, 158, 177
Stonewood (Michigan), 117–120
Stove, wood, 19, 135, 163, 205
Stovewood. *see* Cordwood wall construction
Straightening cordwood walls, 172-173
Straw and clay mortar, 140–143
Straw bales, 135
Styrofoam board, 173
Swedish cordwood buildings, 8–9, 151–155

T

Table for saw, construction of, 74–77, 123
Temporary posts, value of, 52
Thermal mass, for heat storage, 163, 195, 218–219
Thermo-shutters, 164
Turf. *see* Roofing

U

Underfloor radiant heating, 38–39
Used building materials. *see* Recycling

V

Vapor barrier. *see also* Moisture management
    damage to, 196–197
    double wall, use in, 40–41
Ventilation, 163

W

Wall construction. *see* Cordwood wall construction; Post and beam framework
Wall straightening, 172
Water supply, cistern, 100

Windows. *see also* Solar design
    designing log pattern around, 73
    energy efficient, 101
    fire barriers, 198
    framing around, 100, 127
    Las Ventanas Naturales, 147–148
    to open or keep closed, 17
    shutters, insulated, 164
    solar gain from, 162
Wiring. *see* Electrical installation
Wood
    cutting into log ends, 74, 123
    debarking or skinning, best time to, 23–25, 161, 170
    drying time, 22–23, 128, 154–155
    estimating quantity to cut, 24–25
    finishing inside with polyurethane, 41
    log-end shrinkage/expansion, 22–23, 129, 154–155
    for log-ends, 21–22, 27, 133, 152, 160, 170–172
    moisture management, 171–174, 196–197
    moisture absorption of log-ends, 85–86, 154–156
    preservative, use of, 101, 155
    preventing decay or rot, 22–23, 129, 194, 197
    R-value of log-ends, 26, 195, 218–219
    salvage, 105–112, 178, 191
    splitting logs, reduce checking and shrinkage, 27
Woodstoves, 19, 129, 135, 163, 205
Work parties, optimizing, 176, 191
Workshops, 71–74, 131, 140, 155
Wrench, Tony, (round house, Wales), 133–138

About the Authors

George Adkisson lives with his wife, Gwen, in their cordwood home in West Columbia, Texas. (Chapter 18.)

John Carlson and **Scott Carrino** live with their families at the Pompanuck Community, near Cambridge, New York. They co-hosted the 1999 Contintental Cordwood Conference at Pompanuck. (Chapter 24 and 29.)

Val Davidson and her husband Jim live in a beautiful round owner-built cordwood home in Parson, British Columbia. (Chapter 8.)

Dr. Kris Dick, Professional Engineer, has a consulting business, Building Alternatives. Inc., in Anola, Manitoba. Professor A.M. Lansdown is deceased. Together, Dick and Lansdown wrote *Stackwall: How to Build It* (see bibliography.) (Chapter 27.)

Al Fritsch and **Jack Kieffer** work with Appalachia-Science in the Public Interest (ASPI) near Mount Vernon, Kentucky. (Chapter 13.)

Hans Hebel lives with his large family in the Pualafquen Valley near Conaripe, Chile. Cordwood masonry has become a mainstay of an eco-village that is underway in the valley. (Chapter 21.)

Jack Henstridge of Upper Gagetown, New Brunswick, is often referred to as the father of the modern cordwood movement. He is the author of *The ABC's of Cordwood* (see bibliography). (Chapter 5.)

Wayne Higgins lives his wife, Marlys, at Stonewood, their owner-built stackwall-corned home in Calumet, Michigan. (Chapter 16.)

Tom Huber lives with his wife Holly and two young daughters at their homestead near Watervliet, Michigan, where Tom continues to experiment with cordwood masonry. (Chapter 23.)

Geoff Huggins is a retired acoustical engineer. He lives with his partner Louisa Poulin in a cordwood earth-sheltered home they built themselves. (Chapter 12.)

Jim Juczak, with his family, lives in his owner-built two-story 18-sided cordwood home in Adams Center, New York. (Chapter 14.)

Stephen and **Christine Ketter-McDiarmid** live in their beautiful self-built cordwood home in Annapolis, Missouri. (Chapter 25.)

Thomas M. Kwitakowski, deceased, was a fireman and code enforcement officer for the City of Plattsburgh, New York. He built his own cordwood home in the 1980's. (Chapter 28.)

Olle Lind lives with his wife and lots of cockatoos in his owner-built cordwood home in Sweden's "far north, in the woodlands." (Chapter 22.)

Paul Mikalauskas, deceased, built a round cordwood home in Ashland, New Hampshire. Michael Abel built his round cordwood home in Wetherby, Missouri, where he is a licensed electrician and contractor. (Chapter 10.)

Rob Roy is Director of Earthwood Building School, West Chazy, New York. He lives in West Chazy with Jaki, his wife of 30 years, who is also a cordwood co-conspirator. (Chapters 2, 3, 6, 7, 9, 11, 20, 26, 30.)

Larry Schuth is a retired schoolteacher living with his wife, Char, in Hilton, New York. On weekends, and during the summer, they enjoy their owner-built cordwood cabin near Watertown, New York. (Chapter 17.)

Cliff Shockey is the author of *Stackwall Construction: Double Wall Technique* (see bibliography) and lives with his wife Sylvie in their double-wall cordwood home in Vanscoy, Saskatchewan. (Chapter 4.)

Alan Stankevitz was nearing completion of his 16-sided cordwood home in SE Minnesota at press time. He manages the excellent cordwood website, http://www.daycreek.com (Chapter 15.)

Prof. William H. (ìBillî) Tishler, recently retired, was professor of landscape architecture at the University of Wisconsin at Madison. He lives in Madison. (Chapter 1.)

Tony Wrench, with his partner Jane Faith, built their cordwood and cob, earth-sheltered "Round House" at the Brithdir Mawr Community near Newport, Pembrokeshire, Wales. (Chapter 19.)

If you have enjoyed *Cordwood Building: The State of the Art* you might also enjoy other

BOOKS TO BUILD A NEW SOCIETY

Our books provide positive solutions for people who want to make a difference. We specialize in:

**Sustainable Living • Ecological Design and Planning • Natural Building & Appropriate Technology
New Forestry • Environment and Justice • Conscientious Commerce • Progressive Leadership
Educational and Parenting Resources • Resistance and Community • Nonviolence**

For a full list of NSP's titles, please call 1-800-567-6772 or check out our web site at:
www.newsociety.com

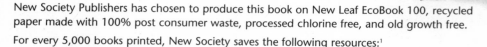

New Society Publishers

ENVIRONMENTAL BENEFITS STATEMENT

New Society Publishers has chosen to produce this book on New Leaf EcoBook 100, recycled paper made with 100% post consumer waste, processed chlorine free, and old growth free.

For every 5,000 books printed, New Society saves the following resources:[1]

53	Trees
4,837	Pounds of Solid Waste
5,322	Gallons of Water
6,942	Kilowatt Hours of Electricity
8,793	Pounds of Greenhouse Gases
38	Pounds of HAPs, VOCs, and AOX Combined
13	Cubic Yards of Landfill Space

[1] Environmental benefits are calculated based on research done by the Environmental Defense Fund and other members of the Paper Task Force who study the environmental impacts of the paper industry.

For more information on this environmental benefits statement, or to inquire about environmentally friendly papers, please contact New Leaf Paper – info@newleafpaper.com Tel: 888 • 989 • 5323.

NEW SOCIETY PUBLISHERS